The Exstrophy–Epispadias Complex

Research Concepts and Clinical Applications

The Exstrophy–Epispadias Complex

Research Concepts and Clinical Applications

Edited by

John P. Gearhart

Johns Hopkins Medical Institutions
Baltimore, Maryland

and

Ranjiv Mathews

SUNY Health Science Center at Syracuse
Syracuse, New York

Kluwer Academic / Plenum Publishers
New York, Boston, Dordrecht, London, Moscow

Library of Congress Cataloging-in-Publication Data

The exstrophy-epispadias complex : research concepts and clinical applications / edited by John P. Gearhart and Ranjiv Mathews.
p. cm.
"Proceedings of the First International Symposium on Exstrophy and Epispadias, held April 9-10, 1997, in Baltimore, Maryland"--T.p. verso.
Includes bibliographical references and index.
ISBN 0-306-46128-5
1. Bladder exstrophy Congresses. 2. Epispadias Congresses. I. Gearhart, John P. II. Mathews, Ranjiv. III. International Symposium on Exstrophy and Epispadias (1st : 1997 : Baltimore, Md.)
[DNLM: 1. Bladder Exstrophy--surgery Congresses. 2. Epispadias--surgery Congresses. 3. Reconstructive Surgical Procedures Congresses. WJ 500 E96 1999]
RD580.E97 1999
617.4'62'0083--dc21
DNLM/DLC
for Library of Congress 99-16120
CIP

Cover figure from, Canning, D. A., Gearhart, J. P., Peppas, D. S., and Jeffs, R. D., "The cephalotrigonal reimplant in bladder neck reconstruction for patients with exstrophy or epispadias," Journal of Urology, **150**: 156, 1993.

Proceedings of the First International Symposium on Exstrophy and Epispadias, held April 9–10, 1997, in Baltimore, Maryland

ISBN 0-306-46128-5

233 Spring Street, New York, N.Y. 10013

10 9 8 7 6 5 4 3 2 1

A C.I.P. record for this book is available from the Library of Congress.

Printed in the United States of America

DEDICATION

This book is dedicated to the following people: Madison Pryor, Ph.D. who taught me that the study of biology is the basis of medicine; to Lonnie W. Howerton, M.D. who taught me that urology is the pinnacle of surgical sub-specialities; to Hiram C. Polk, M.D. who taught me to ask questions, and more importantly to find answers; to Roy Witherington, M.D. who taught me the skills and science of urology; to Mr. J. Herbert Johnson, MB, FRCS, who taught me the craft of pediatric urology; to Robert D. Jeffs, M.D. who taught me the skills and science of reconstructive pediatric urology; and to Patrick C. Walsh, M.D. for his mentorship and constant dedication to discovery and achievement.

PREFACE

The undertaking of the treatment of an infant born with bladder exstrophy is one of the most weighty responsibilities that can fall upon the shoulders of the reconstructive surgeon. The modern treatment of a child born with bladder exstrophy began in the mid-1970's with the widespread application of staged reconstruction. This approach has consistently yielded very good results in several series. However, as in all serious congenital birth defects, there is certainly room for advancement.

Issues such as the routine use of osteotomy, timing and type of epispadias repair, combining bladder closure with epispadias repair, the approach to the small bladder, and the management of a failed exstrophy still remain. This National Institute of Health/National Kidney Foundation/Johns Hopkins-sponsored seminar was an attempt to bring experts in the field of pediatric orthopedic surgery, pediatric urology, pediatric surgery, adult urology, and basic science together to share their experiences in an attempt to foster new clinical and basic science research communications between the participants. If these collaborations result, then this first international meeting will have been successful.

The editors would like to thank all of the contributors for their timely and complete submissions.

John P. Gearhart, M.D.
Ranjiv Mathews, M.D.

CONTENTS

1

THE EMBRYOLOGY AND EPIDEMIOLOGY OF BLADDER EXSTROPHY

Bruce Slaughenhoupt

Division of Urology
University of Louisville School of Medicine
Louisville, Kentucky 40292

INCIDENCE

In 1987, the International Clearing House for Birth Defects estimated that classic bladder exstrophy occurred at a rate of 1 in 30,000 live births.[1] In a more recent study published in 1994, the University of Pittsburgh estimated classic bladder exstrophy to occur once in 25,000 live births in the Pennsylvania, Virginia, Maryland, and Washington, D.C. areas.[2] The University Hospital in Lund, Sweden recently cited an incidence of 1 in 33,500 live births.[3]

The spectrum of bladder exstrophy and epispadias complex occurs much more commonly in males than in females. In 1973, Bennett reported a 2.3 to 1 ratio[4] and in 1980, Ives reported a 6 to 1 male to female ratio.[5] Shapiro has also reported that a person with exstrophy or epispadias has a 1 in 70 chance of having a child with bladder exstrophy.[6] Most of the children reported in this study were children born to women with bladder exstrophy. Thus far, there have been very few reports of men with bladder exstrophy who have fathered children. Most of these cases remain anecdotal.

EXTERNAL APPEARANCE

I would like to briefly describe the external anatomic appearance of a young child born with classic bladder exstrophy for those who do not care for these patients in the clinical setting. Classic bladder exstrophy is an infraumbilical defect in which the bladder's internal mucosal surface is exposed to the outside environment. The abdominal wall fascial defect is limited inferiorly by the intersymphyseal band. The urethral plate is exposed on its dorsal surface and is surrounded by dorsal chordee. This causes the phallus to be drawn up against the abdominal wall and the exposed bladder. The penis is broad and

The Exstrophy–Epispadias Complex, edited by Gearhart and Mathews.
Kluwer Academic / Plenum Publishers, New York, 1999.

foreshortened. In the female, the vagina and urethra are short, the clitoris is bifid and the labia are divergent. In both sexes, the symphysis pubis is widened and has an outward rotation of the innominate bones at the sacroiliac joints. The anus is anteriorly displaced. Gearhart and Peppas have shown that males born with classic bladder exstrophy have an 83% incidence of indirect inguinal herniae being present.[7]

NORMAL EMBRYOLOGIC GENITAL DEVELOPMENT

I would like to briefly review normal embryonic development prior to discussing the theories of the abnormal embryonic development which may lead to classic bladder exstrophy. At three weeks gestation, the bilaminar cloacal membrane normally undergoes mesodermal ingrowth and caudal migration. That is, mesodermal tissue migrates from a lateral to medial direction as well as from a cranial to caudal direction. The mesodermal tissue then leads to formation of the abdominal wall and bony pelvic anatomy. At the conclusion of the mesenchymal infiltration of the cloaca, the distal most aspect of the mesenchyme leads to labioscrotal swellings and formation of the genital tubercle.

At 4 weeks gestation, Tourneux's folds advance in a cranial-to-caudal direction and Rathke's plicae grow toward each other in a lateral-to-medial direction. Therefore by 7 weeks gestation, the urorectal septum has migrated completely and merged with the cloacal membrane to form the perineal body.[8] It is believed that the thin, bilaminar cloacal membrane remnant ruptures due to the pressure of mesonephric urine. The cloacal membrane ruptures anteriorly and leads to the opening of the urogenital sinus while the posterior aspect opens into the rectum.

EMBRYOLOGY OF BLADDER EXSTROPHY

During the past century, there have been several theories proposed to explain the etiology of bladder exstrophy. In 1869, Wood proposed that the defect was due to ulceration of the anterior abdominal wall and symphysis pubis. This was then thought to lead to necrosis and separation of the abdominal wall and pubis. It was actually suggested at the time that syphilis infection might lead to bladder exstrophy.[9]

In 1924, von Geldern summarized the views of several contemporaries who ascribed bladder exstrophy as a developmental arrest in a normal embryologic phase through which all fetuses pass. This has since been found to be at fault since fetuses that die of other causes, do not also have findings consistent with bladder exstrophy.[10]

In 1901, an article in *JAMA* supported the Berstung theory.[11] This proposed that bladder outlet obstruction led to rupture of the anterior bladder wall and abdominal wall. It was also suggested that prior to its rupture, the distended bladder caused lack of fusion of the bony pelvis and, therefore, separation of the pubic symphysis. It was proposed that after rupturing, the bladder wall fused with the abdominal wall and healed together. It is now felt that bladder outlet obstruction does not lead to classic bladder exstrophy. In fact, many boys are born each year with bladder outlet obstruction from posterior urethral valves. Their clinical picture is very different from those of children born with bladder exstrophy and often includes severe hydronephrosis, dilated ureters and an enlarged bladder.

In 1952, Patten and Barry theorized that bladder exstrophy and epispadias were due to misplacement of the embryologic genital tubercles.[12] They felt that the genital tubercles in these fetuses were in an abnormally low position. Therefore, midline abdominal wall fusion developed below the tubercles but not above them. When the genital tubercles arise below the urogenital orifice, it is theorized that they cannot augment the abdominal wall and urethral development.[11] This concept has fallen out of favor in recent years because, according to its theory, epispadias should be more likely to occur than classic bladder exstrophy. However statistically, simple epispadias is much rarer than classic bladder exstrophy. Therefore, this theory is subsequently felt to not be accurate.

In 1962, Marshall and Muecke published their paper proposing a very reasonable theory concerning the etiology of bladder exstrophy.[13] They suggested that an abnormality of the cloacal membrane prevented it from allowing mesodermal ingrowth, and subsequent abdominal wall development. They felt the membrane remained bilaminar and was thinner than it should be. It was then felt that the membrane ruptured under the force of urine produced by primitive kidneys. They also proposed that the less mesodermal ingrowth which occurred at the time of rupture, the more severe the defect would be. That is, if very little mesoderm was added to the cloacal membrane, the child would be born with cloacal exstrophy. If more mesoderm migrated, the child would be born with classic bladder exstrophy. And lastly, if there was a greater amount of tissue which augmented the cloacal membrane, the child would simply be born with epispadias.

Muecke then published an article two years later and reported the use of an animal model to demonstrate the effects of the non-regressing cloacal membrane.[14] He used fertilized chick eggs which had been incubated for two of their 20-day incubation period. A 2 mm incision was made in the region of the chick cloacal membrane. A small wedge of millipore plastic was then inserted into the incision. The eggshell was then resealed and incubation continued. Muecke performed this procedure on 54-dozen chick eggs. None of the chicks of the first 40 dozen survived. Thirteen percent of Muecke's chick embryos survived to hatching and the great majority of them had a caudal defect similar to cloacal or bladder exstrophy. Muecke then concluded that the basic embryologic defect which leads to classic bladder exstrophy occurs early in embryogenesis in the form of an overly developed, non-regressing cloacal membrane. He theorized that this acted as a mechanical barrier to mesodermal migration and subsequent abdominal wall and bladder development.

In 1985, Mitchell, et al, theorized that exstrophy was not due to a persistent cloacal membrane, but rather due to its premature rupture and subsequent eventration of the cloaca.[15] Like Muecke, they also used chicks in an animal model. However, rather than insert a wedge of plastic to mimic a persistent cloacal membrane, they used a laser to incise the membrane in the lower abdominal wall and created a hernia defect. The chicks then incubated through gestation and were born with cloacal exstrophy. They, therefore, concluded that the embryologic defect that most likely leads to exstrophy is not a persistent cloacal membrane, but rather an untimely, premature rupture of the membrane. This is the theory held by most at the present time to explain the etiology of classic bladder and cloacal exstrophy.

REFERENCES

1. Epidemiology of bladder exstrophy and epispadias: A communication from the International Clearinghouse for Birth Defects Monitoring Systems. *Teratology*, 36:221, 1987.
2. Yang P, Khoury MJ, Stewart WF, et al: Comparative epidemiology of selected midline congenital abnormalities. *Genetic Epidemiology*, 11:141–54, 1994.
3. Kockum CC, Hansson E, and Stenberg A, et al.: Bladder exstrophy in Sweden, a longterm folllow-up study, *Eur J Ped Surg*, 6:208–11, August 1996.

4. Bennett AH: Exstrophy of the bladder treated by ureterosigmoidostomies, long term evaluation. *Urology*, 2:165, 1973.
5. Ives E, Coffey R, Carter CO: A family study of bladder exstrophy, *J Med Genetics*, 17:139–41. 1980.
6. Shapiro E, Lepor H, and Jeffs RD: Inheritance of the exstrophy-epispadias complex. *J Urol*, 132:308, 1984.
7. Connolly JA, Peppas DS, Jeffs, RD, Gearhart JP: The prevalence and repair of inguinal hernias in children with bladder exstrophy. *J Urol*, 154(5): 1900, 1995.
8. Stephens FD, Smith ED: Anorectal malformations in children. Chicago, Year Book Medical Publishers, 1971.
9. Wood J: On fission and extroversion of the bladder with epispadias. *Tr Medico-Chir*, 52:85, 1869.
10. von Geldern CE: The etiology of exstrophy of the bladder. *Arch Surg*, 8:61, 1924.
11. Connell FG: Exstrophy of the bladder: *JAMA*, 36:637, 1901.
12. Patten JF, Barry A: The genesis of exstrophy of the bladder and epispadias. *Am J Anat*, 90:35, 1952.
13. Marshall VF, Muecke EC: Variations in exstrophy of the bladder. *J Urol*, 88:766, 1962.
14. Muecke EC: The role of the cloacal membrane in exstrophy: the first successful experimental study. *J Urol*, 92:659, 1964.
15. Thomalla JV, Rudolph RA, Rink RC, and Mitchell ME: Induction of cloacal exstrophy in the chick embryo using the CO_2 laser. *J Urol*, 134:991–5, 1985.

2

PRENATAL DIAGNOSIS OF BLADDER AND CLOACAL EXSTROPHY AND RELATED CONDITIONS

Roger C. Sanders

Department of Radiology
University of Maryland

Bladder exstrophy (Barth et al., 1990; Bronshtein et al., 1993; Gearhart et al., 1995; Jaffe et al., 1990; Khandelwal et al., 1996; Messelink et al., 1994; Pinette et al., 1996) and its related cousin hypospadias (Sides et al., 1996), are among the most challenging entities that can be identified before birth with ultrasound. Hypospadias is the most common of these lesions, but is most difficult to diagnose. A series of six cases and two case reports of the prenatal diagnosis of hypospadias exist (Sides et al., 1996). Typically, the penis is short and it is often curved. The distal end of the phallus is widened forming a club shape. Since, color flow Doppler records flow movement, whether of blood or urine, color flow Doppler permits the visualization of urination, which is seen as an intermittent jet. Normally, jets of urine are seen coming from the tip of the penis but in hypospadias, the urine originates from a more proximal and inferior site than the tip of the penis. The penis has to be examined in a sagittal profile view for a decision as to the source if jet effect to be made with confidence.

Hypospadias is associated with a number of syndromes, including Opitz BB syndrome, Opitz G syndrome and the Smith Lemli Opitz syndrome. It is also seen with some chromosomal anomalies, such as triploidy, 4P syndrome and 13Q syndrome. Particular attention to the configuration of the penis is often the result of finding of some other anomaly such as cleft lip or club feet, and therefore, the fetus as a whole is submitted to a very detailed survey and hypospadias is then discovered. Isolated hypospadias detection is a much greater challenge, but should be possible when there is an appropriate family history.

Bladder exstrophy is somewhat easier to diagnose because the bladder is a structure that is routinely sought during a standard obstetrical sonogram performed for other reasons. The absence of the bladder is a hallmark of bladder exstrophy. Eight individual case reports, in which the diagnosis has been made antenatally, have been reported (Mirk et al., 1986; Barth et al,, 1990; Jaffe et al., 1990; Gearhart et al., 1989) and a large retrospective

The Exstrophy–Epispadias Complex, edited by Gearhart and Mathews.
Kluwer Academic / Plenum Publishers, New York, 1999.

series of 25 fetuses with bladder exstrophy has been reported (Gearhart et al., 1995). The diagnosis of bladder exstrophy could be made retrospectively in 29 prenatal studies from 17 pregnancies. The time of the fetal ultrasound varied from 14–36 weeks gestation (mean 23). The diagnosis of bladder exstrophy, however, was made before delivery in only three cases. Five abnormal ultrasonic features associated with bladder exstrophy were identified: 1) the bladder was not visualized on ultrasound in 12 of the 17 cases (71 percent); 2) a lower abdominal bulge representing the exstrophied bladder was seen in 8 of the 17 cases (47 percent); 3) a small penis with anteriorly displaced scrotum was identified in 8 out of 14 males (57 percent); 4) the umbilical insertion was low-set in 5 of 17 cases (29 percent); and 5) abnormal widening of the iliac crest was seen in 3 of the 17 cases (18 percent). An absent bladder is a consistent finding which is always seen, however the changes in location of the genitalia and cord insertion are subtle and only visualized if the fetus is examined in the sagittal alignment with the fetus supine. Mounding of the bladder remnant is highly variable. In some instances, this is very obvious, and in others the exstrophied bladder has a flat surface. Secondary obstructive changes in the kidneys may occasionally occur with bladder exstrophy.

The iliacs normally form a right angle to one another. In bladder exstrophy, this angle is widened to about 110 degrees. This change in iliac crest angle is relatively easy to recognize in utero because a view of the lumbosacral spine, obtained to rule out myelomeningocele is routine. Recently, a large series of fetuses have had the iliac angle analyzed in relation to Down's syndrome. A widened iliac angle is also a sign of Down's syndrome (Kliewer et al., 1996). These normal standards for iliac angle can also be used for the detection of bladder exstrophy.

The differential diagnosis of bladder exstrophy includes other abdominal wall malformations such as omphalocele, which lies at a higher level since it is associated with the cord insertion. In addition, the bladder is present with omphalocele. Gastroschisis, another abdominal wall malformation, arises in the lower abdomen on the right. Gut occasionally, including the bladder or stomach, lies outside the fetal abdomen with the gastroschisis, exiting through a hole in the right iliac region. As a rule, the bladder is in its usual location, so these conditions are not easily confused with exstrophy. Absence of the bladder may occasionally be a normal finding in utero, although as a rule, the fetus does not completely empty its bladder when it voids. Other causes of absent bladder that should be considered include renal agenesis and any cause of absent urine output, such as bilateral multicystic dysplastic kidney or severe infantile polycystic disease.

Although the absence of the bladder should make the diagnosis of bladder exstrophy relatively easy to make, it is often missed. In the series by Gearhart et al., in 1995, only 3 out of 25 prenatally seen bladder exstrophy cases were recognized prenatally. In retrospect, 17 had features consistent with bladder exstrophy, such as absent bladder. Fourteen of these cases were male and 3 female. It is easier to make this diagnosis in males because of the genitalia changes. The fetal bladder is visible by 11–12 weeks. It becomes more obvious as the pregnancy progresses. With modern equipment, the kidneys can be seen at 14–15 weeks.

Cloacal exstrophy (Bruch et al., 1996; Carr et al., 1994; Chitrit et al., 1993; Cilento et al., 1994; Kutzner et al., 1988; Mandell et al., 1995; Meizner et al., 1995; Nishi et al., 1988; O'Leary et al., 1995; Richards et al., 1992; Winderi et al., 1996; Zimmer et al., 1996) is a syndrome with many obvious ultrasonic features. At least 17 single or double case reports have been published of this entity discovered in utero. Although the many ultrasonic features are obvious, it is sometimes difficult to combine them all to make the right diagnosis. With the usual open type of cloacal exstrophy, there is absence of the fetal bladder. There is a large infraumbilical anterior midline defect with a massive protruding omphalocele. The thorax is narrowed and the spine is kyphoscoliotic. There may well be caudal regression. A sacral meningocele is often present with possible secondary changes in the cranium of the Arnold Chiari Type I phenomenon. Bilateral club foot is usually seen. Ascitic fluid is often present. Again, the iliac crest is widened. Recently, the anus has been recognized by ultrasound. Normally, there is a small echogenic spot surrounded by an echopenic area on a transverse through view at the level of the external genitalia. This finding is absent in cloacal exstrophy since imperforate anus is a typical feature of the syndrome. Obstruction of the kidneys and ureters is also commonplace with perhaps, the development of multicystic dysplastic kidney. The syndrome appears to be much more common in monochorionic twins than expected. A subvariant is the OEIS complex (omphalocele, vesical exstrophy, imperforate anus and spinal abnormalities) and this is the form which occurs most often in twins (Chitrit et al., 1993; Lee et al., 1997; Sherer et al., 1992).

Many of the milder cases are confused with omphalocele since caudal regression and spinal changes are not easy to recognize. Severe examples of the entity are easily confused with two other similar syndromes. The body stalk anomaly is a syndrome in which the cord is exceptionally short. Cord tension is thought to be the reason why much of the abdominal contents are exteriorized. There is myelomeningocele with secondary changes of Arnold Chiari type 1 in the cranium. There is usually a markedly distorted spine which may be shortened with caudal regression and is almost always curved and scoliotic. Another similar syndrome is the limb body wall complex, or cillosomia. In this entity, disruption of the amnion at approximately 6–8 weeks with subsequent amniotic band formation, is thought to be responsible for many malformations. Almost always, one or more of the limbs is absent. The spine is kyphoscoliotic and there is usually caudal regression. A spina bifida is again a frequent feature with secondary cranial changes of the Arnold Chiari Type I malformation. The abdominal wall is disrupted with most of the contents of the abdomen lying anterior to the abdominal wall. Ectopia cordis is a common secondary malformation. Both body stalk anomaly and limb body wall complex are lethal anomalies.

The key in distinguishing cloacal exstrophy, which is salvageable from lethal body stalk anomaly or limb body wall complex, lies in close examination of the extent of the caudal regression and seeing whether all limbs are present. As a rule, with limb body wall complex, there is little or no amniotic fluid. In both conditions, the large myelomeningocele can appear to be a sac containing amniotic fluid. Ascites surrounding the contents of the large omphalocele can also be mistaken for amniotic fluid.

Hypospadias, bladder exstrophy and cloacal exstrophy represent rare entities which are difficult to find with ultrasound. Experienced ultrasound practitioners should be able to find all three problems when a 16–18 week standard obstetrical sonogram is performed. Knowledge of the nature of the anomaly is very helpful to parents in deciding whether to continue the pregnancy. It is standard practice to have a pediatric urologist talk to the patient about anomalies when they are discovered They counsel them on the type of surgery, duration of time in hospital, delivery site and time of delivery. Patients with such anomalies have a multitude of questions that only the pediatric urologist can properly answer.

REFERENCES

1. Barth RA, Filly RA, Sondheimer FK: Prenatal sonographic findings in bladder exstrophy. J Ultrasound Med 1990: 9 359–361.
2. Bronshtein M, Bar-Hava I., Blumenfeld Z.: Differential diagnosis of the nonvisualized fetal urinary bladder by transvaginal sonography in the early second trimester. Obstet Gynecol 1993; 82:490–493.

3. Bruch SW, Adzick NS, Goldstein RB, Harrison MR. Challenging the embryogenesis of cloacal exstrophy. J Pediatr Surg 1996; 31:768–770.
4. Carr MC, Benacerraf BR, Mandell J. Prenatal diagnosis of an XY fetus with aphallia and cloacal exstrophy variant. J Ultrasound Med 1994;13:323–325.
5. Chitrit Y, Zom B, Filidori M, Robert E, Chasseray JE. Cloacal exstrophy in monozygotic twins detected through antenatal ultrasound scanning. J Clin Ultrasound 1993; 21:339–342.
6. Cilento Jr BG, Benacerraf BR, Mandell J. Prenatal diagnosis of cloacal malformation. Urology 1994; 43:386–388.
7. Gearhart JP, Ben-Chaim J, Jeffs RD, Sanders RC. Criteria for the prenatal diagnosis of classic bladder exstrophy. Obstet Gynecol 1995. 85:961–964.
8. Gearhart JP, Jeffs RD, State of the art reconstructive surgery for bladder exstrophy at the Johns Hopkins Hospital. Am J Dis Child 1989; 143:1475–1479.
9. Jaffe R, Schoenfeld A, Ovadia J. Sonographic findings in the prenatal diagnosis of bladder exstrophy. Am J Obstet Gynecol 1990; 162:675–678.
10. Khandelwal M, Coyman RC, Barahona O, Schmitt R, Reece EA. Early prenatal diagnosis of bladder exstrophy: Case report and review of the literature. Fetal Diagn Ther 1996; 11:146–149.
11. Kleiwer MA, Hertberg BS, Freed KS, DeLong DM, Kay HH, Jordan SG, Peters-Brown TL, McNally PJ. Dysmorphologic features of the fetal pelvis in Down syndrome: Prenatal sonographic depiction and diagnostic implications of the iliac angle. Radiology 1996; 201:681–684.
12. Kutzner DK, Wilson WG, Hogge WA. OEIS complex (Cloacal exstrophy) prenatal diagnosis in the second trimester. Prenat Diagn 1988; 8:247–253.
13. Lee DH, Sanders RC, Meyers CM, Wulfsberg EA, Sun CCJ. The OEIS complex (omphalocele-exstrophy-imperforate anus-spinal defects) in monozygotic twins. Proceedings: United States and Canadian Academy of Pathology 1997; 10:6A.
14. Mandell J, Bromley B, Peters CA, Benacerraf BR. Prenatal sonographic detection of genital malformations J Urol 1995; 153:1994–1996.
15. Meizner I, Levy A, Barnhard Y. Cloacal exstrophy sequence: An exceptional ultrasound diagnosis. Obstet Gynecol 1995: 86:446–450.
16. Messelink EJ, Aronson DC, Knuist M, Heij HA, Vos A. Four cases of bladder exstrophy in two families J Med Genet 1994; 31:490–492.
17. Mirk P, Calisti A, Fileni A. Prenatal sonographic diagnosis of bladder exstrophy. J Ultrasound Med 1986; 5:291–293.
18. Nishi T, Yamoto M, Nakano R. Prenatal diagnosis of exstrophy of the cloaca with ectopia cordis. Asia Oceania J Obstet Gynecol 1988; 14:213–217.
19. O'Leary r, Davies P, Challis D, Bury G. Closed bladder exstrophy (pseudoexstrophy) associated with mesodermal anomalies. J Ultrasound Med 1995; 14:311–314.
20. Pinette MG, Pan YQ, Pinette SG, Stubblefield PG, Blackstone J. Prenatal diagnosis of fetal bladder and cloacal exstrophy by ultrasound. A report of three cases. J Reprod Med 1996; 41:132–134.
21. Richards DS, Langham Jr MR, Mahaffey SM. The prenatal ultrasonographic diagnosis of cloacal exstrophy. J Ultrasound Med 1992; 11:507–510.
22. Sherer DM, Abramowicz JS, Smith SA, Cusson CL, Metlay LA, Woods Jr JR, Normal lung structure and pulmonary function in a twin with agenesis of the cloacal membrane and persistent severe oligohydramnios. J Maternal-Fetal Med 1992; 1:24–28.
23. Sides D, Goldstein RB, Baskin L, Kleiner BC. Prenatal diagnosis of hypospadias. J Ultrasound Med 1996: 15:741–746.
24. Winderi LM, Patel V. Hydrometrocolpos associated with uterovaginal duplication: The prenatal sonographic findings in a case of persistent cloaca. J Diagn Med Sonogr 1996; 12:136–140.
25. Zimmer EZ, Bronshtein M. Early sonographic diagnosis of fetal midline disruption syndromes. Prenat Diagn 1996; 16:65–69.

3

ANATOMY AND BIOMECHANICS OF THE PELVIS IN EXSTROPHY

Mihir M. Jani, Paul D. Sponseller, John P. Gearhart, A. Valdevit, and E. Y.-S. Chao

Departments of Orthopedics and Urology
The Johns Hopkins School of Medicine
Baltimore, Maryland

I. ANATOMY OF THE PELVIS IN CLASSIC AND CLOACAL EXSTROPHY

The Exstrophy Complex includes anomalies ranging from epispadias to classic to cloacal exstrophy. Exstrophy of the bladder occurs due to abnormal anterior rupture of the cloacal membrane early in the embryonic period. Mesenchymal ingrowth into the abdominal wall is inhibited. Since the pelvis is derived from sclerotomal components of the mesenchyme, one may hypothesize that the development of the entire pelvis may be affected by this altered migration of mesenchyme. It is important to understand the patho-anatomy of the pelvis in exstrophy, for planning of appropriate soft-tissue and bony correction in this disorder. However, there are very few specimen dissections reported, and no comprehensive analysis of pelvic shape in either classic or cloacal exstrophy. Therefore, analysis of computed tomograms of patients with classic and cloacal exstrophy was carried out to better define the differences in the shape of the pelvis between these patients and those without congenital anomaly. We attempted to answer the question of where the pelvis in exstrophy was rotated, to quantify the malrotation, and to determine whether there was an actual deficiency of growth. An analysis of the rotation and the spacing of the hips was also made. In addition, we wished to correlate these findings with the clinical picture of patients' gait, which is commonly described as "out-toeing" or "waddling."

Method of Analysis

The bony pelvis is comprised of three bones in addition to the sacrum: the ilium, ischium and pubis, which are joined at the triradiate cartilage of the acetabulum. The ilium is posterior, while the ischium and pubis are anterior to it. For the purpose of this analysis, the pelvis was conceptualized as being composed of an anterior (ischio-pubic) segment and a

The Exstrophy–Epispadias Complex, edited by Gearhart and Mathews.
Kluwer Academic / Plenum Publishers, New York, 1999.

posterior (iliac) segment, joined at the triradiate cartilage. Computed tomographic (CT) scans were obtained of twenty-five patients who had not had an osteotomy for exstrophy: 19 had classic exstrophy and 9. cloacal exstrophy. Clinical exams were performed on each patient, assessing the gait , strength and rotation. Controls were obtained from 36 age-matched patients who underwent pelvic CT for other reasons and had no congenital anomaly. Most were done for trauma, and those with pelvic fractures or dislocation were excluded.

On the CT images, parameters measure included the length and angulation of the anterior and posterior segments, as well as the diastasis (spread) of the pubis. The transverse orientation (version) of the acetabulum was also measured, as well as the increased distance between the two hips, which has special biomechanical significance. The anteversion of the femur, or the orientation of the proximal and distal joints in the transverse plane, was also measured in order to determine if any compensatory growth occurred in response to the primary, acetabular change in orientation.

Clinical exams were performed on each patient. In order to study all of the components of the limb which determine the final rotation of the knees and the feet during gait, we measured the internal and external rotation of the hips in extension, as well as the thigh-foot angle (for tibial torsion) and the foot progression angle, or the angle of the foot during gait. The Trendelenburg test was used as a measure of gluteal muscle strength.

Patients with simple epispadias were not examined by CT scan, since the use of radiation could not be justified for a condition which does not require osteotomy. Plain radiographs of the pelvis were routinely available, usually as part of an intravenous pyelogram. These were analyzed for diastasis of the symphysis.

Findings

Rotational Abnormalities. On analysis of the CT data, it was shown that the not only was the anterior portion of the pelvis abnormal, but also the posterior segment was affected as well, with a mean external rotation in classic exstrophy patients of 12 degrees on each side (Figure 1). This was greater in the patients with cloacal exstrophy. The degree of external rotation tended to decrease with time, however. The anterior segment was further externally rotated 6 degrees beyond the posterior segment in the classic exstrophy patients, and showed little tendency to correct with time (Figure 1). The acetabulum was

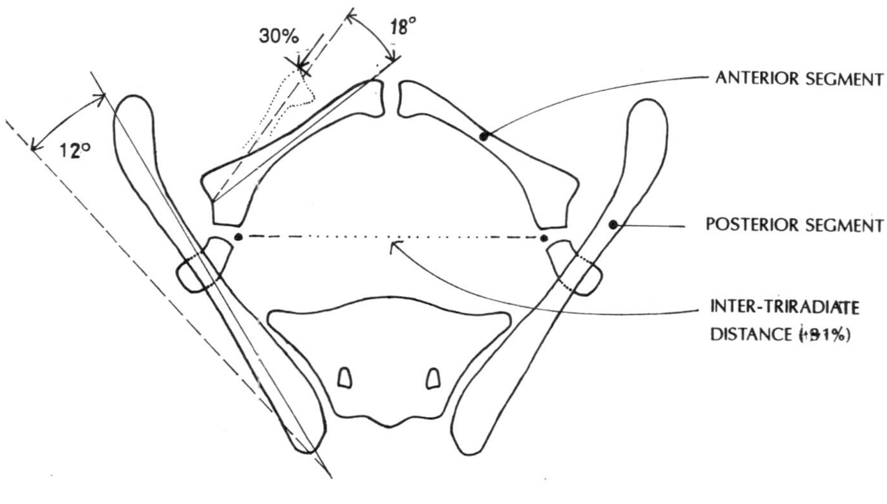

Figure 1. Diagram of the pelvis showing the changes in the anatomy of the pelvis in classical exstrophy.

retroverted, or rotated externally, on each side, but there was no measurable change in the femur in response to this

Size and Shape of Segments. The size of the posterior segment, that is, of the iliac bone, was equal to that of age-matched controls. The anterior segment, however, showed a very interesting and significant decrease in length. In classic exstrophy, this length was 30% less than controls (Figure 1) . Not only were the bones malrotated, but there was also a deficiency of their length anteriorly. In cloacal patients, the anterior deficiency was often much more pronounced.

The diastasis of the pubic symphysis increased linearly in patients with classic exstrophy, from a mean of 4 centimeters at birth to 8 centimeters at age ten, to 14 centimeters at skeletal maturity. The patients with epispadias also had a diastasis of the symphysis pubis, a fact which is not commonly appreciated. By contrast, the normal width of the symphysis in persons without exstrophy or epispadias is essentially constant throughout life, at 0.7 to 1 cm.

Many other malformations were seen in the group with cloacal exstrophy. There was a 25% incidence of hip dislocation. Many patients had malformation of the sacroiliac joint, in which it was underdeveloped and functioned as a hinge joint instead of a strong junction with almost no detectable motion, as in the normal state.

Clinical Exams. The patients with classic exstrophy walked with an external placement of the feet up until approximately school age, after which time the angle began to correct. There were no significant abnormalities in the femur or the tibia. There was no evidence of hip muscle weakness, and all patients were able to participate in sports and physical activities normally.

Summary. The event which occurs early in embryonic life in exstrophy affects not only the genitourinary system but also the migration and formation of the bony pelvis. Specifically, the following have been shown[9]:

- The posterior (iliac) segments are externally rotated a mean of 12 degrees but correct partially with time.
- The anterior (ischiopubic) segments are further externally rotated.
- The anterior segment is shortened by 30%.
- The pubic diastasis increases linearly with age.
- The acetabulum is retroverted.
- The patients with cloacal exstrophy have more severe degrees of all of the above findings. They also may have malformations such as dislocated hips and abnormal sacroiliac joints.

Changes in the pelvis in exstrophy are summarized in Figure 1. They are important to understand in designing an osteotomy to correct the pelvis in order to help with genitourinary reconstruction. They also have implications for the forces about the hip, which may determine its long-term fate.

II. BIOMECHANICAL ANALYSIS OF HIPS IN ADULTS WITH BLADDER EXSTROPHY

Several types of pelvic osteotomies have been developed to decrease the stress on the initial exstrophy closure. Indications for pelvic osteotomy so far have been to lower rates of wound dehiscence and to help obtain a more secure and better genitourinary reconstruction.[1,6] However, pelvic osteotomies have not been shown to improve hip function or gait in children born with bladder exstrophy[11]. The natural history of the hip joint with a congenital, open pelvic ring has not been studied. O'Phalen proposed that the body weight in exstrophy patients is supported by the posterosuperior portion of the femoral heads and that early degenerative changes may result if the pelvis was not corrected[6]. Abnormal peak stresses caused by elevated joint loads can lead to failure of the articular cartilage and development of early osteoarthritis. We studied older patients with uncorrected pelvic deformity of bladder exstrophy in order to determine the relative hip joint force and stress and their effects on the clinical status of the hip.

Method of Analysis

We studied 14 adult patients, with a mean age of 32 years (range 16 to 52 years), who were born with classic bladder exstrophy. Twelve patients did not have pelvic osteotomies at exstrophy closure whereas two patients had osteotomies with essentially no maintenance of correction.

Biomechanical analysis was based on Legal's model of the single-stance phase of gait, the phase at which the maximum load and stress is exerted on the hip joint. The model assumes a static equilibrium between hip abductors and partial body weight around the hip during a single leg stance. Anterior-posterior (AP) pelvic radiographs were used to determine reference points and to make appropriate measurements according to Legal (1987) (Figure 2)[5]. Pubic symphyseal diastasis was determined by measuring the distance between the two most medial points of the pubic rami. In order to compare results among patients and controls, joint force and stress (force/unit area) were calculated relative to partial body weight, which is the joint force or stress divided by the 5/6 of the total body weight since the supporting limb is approximately 1/6 of the total body weight.

In order to correlate theoretical models with joint status *in vivo* in mid-adulthood, hips were scored for degenerative joint disease (DJD) from the same anterior-posterior radiographs used in biomechanical analyses, and clinical hip ratings were obtained.

Patients were matched for age and gender with 14 controls (14 hips), and joint force and stress relative to partial body weight were calculated as above. Statistical analysis was done using matched-pairs and Student t tests. The level of significance was $p < 0.05$.

Findings

The results of biomechanical analysis are summarized in Table 1.

Based on Legal's method, the force and stress on the hip joint are increased in adult exstrophy patients with uncorrected pelvic deformity. Changes in several parameters between exstrophy patients and controls contribute to this increase. In the adult exstrophy patients, the significant increase in intersymphyseal diastasis correlates linearly with an increase in the distance from the body midline to the femoral head (*d6*). This correlates with an almost linear rise in joint load and stress because of an increase in the moment arm through the femoral head created by the partial body weight acting at the body's center of gravity.[5] In addition, the significant decreases in t' and CE angle contribute to increased load and stress on the hip joint as well.[5]

There appear to be two groups of patients in this study: the majority had CE angles in the normal to borderline range, and no significant evidence of degenerative change. These patients had asymptomatic hips and low DJD scores despite having increased load

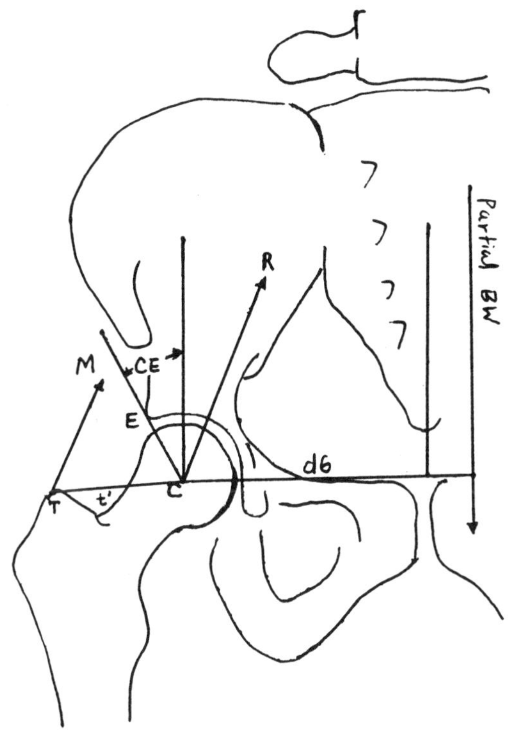

Landmarks:

T - point where horizontal surface of the greater trochanter meets its superiolateral border

C - center of the femoral head

E - lateral edge of the acetabulum

Measurements:

CE angle - angle between vertical through point C and line CE

d6 - perpendicular distance from body midline to point C

t' - distance between points C and T

Other Parameters:

Vector **M** - calculated vector of hip abductor force in single leg stance

Vector **R** - calculated vector of joint force in single leg stance

Vector **Partial BW**- vector of 5/6 total body weight through body center of gravity

Figure 2. Diagram of anterior-posterior view of the pelvis illustrating key landmarks and measurements used in the calculations of relative joint force and stress according to Legal (1987).

Table 1.

	Exstrophy pts.	Controls	P value
Relative joint force	4.24 ± 0.91	3.04 ± 0.26	$p < 0.01$
Relative joint stress (cm^{-2})	0.298 ± 0.18	0.177 ± 0.035	$p < 0.05$
d6 (cm)	12.35 ± 1.05	10.31 ± 0.70	$p < 0.001$
t' (cm)	5.21 ± 1.24	6.52 ± 0.80	$p < 0.02$
CE angle (degrees)	25.8 ± 9.93	33.3 ± 7.73	$p < 0.05$

d6 - distance from body midline to center of femoral head.
t' - distance from greater trochanter to center of femoral head.
CE angle - angle between acetabular edge and center of femoral head.

and stress on the hip joints. However, the second group of hips showed severe hip dysplasia and abnormally small CE angles. In this group, our second oldest patient (JH) had a CE angle of zero degrees (normal is > 20), a low clinical hip score, and significant osteoarthritis in both hips (see Figure 3) which was associated with the highest calculated stress (0.857 cm^{-2}). Abnormal loads and stress on a joint are associated with accelerated onset of osteoarthritis. Given the increased load and stress in the hips of the exstrophy patients, one would expect an earlier onset of degenerative joint disease as shown by JH. Longer follow-up of the younger exstrophy patients in the study may reveal early degenerative joint changes.

The primary indication for pelvic osteotomy presently is to achieve a more secure genitourinary repair. However, pelvic osteotomy may serve to improve hip biomechanics in adult life by restoring the pelvis to as close to its normal anatomy as possible. Pelvic

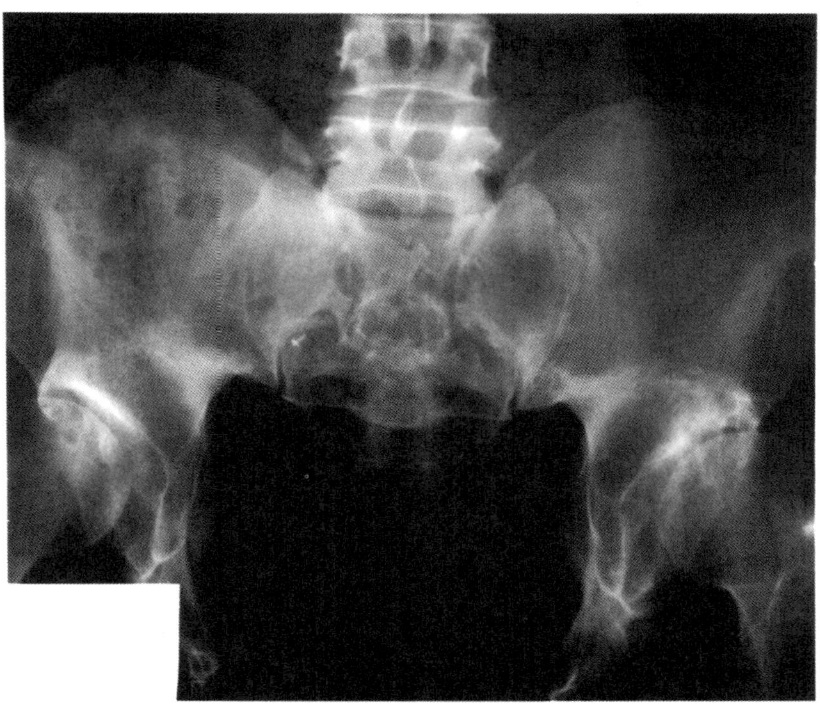

Figure 3. Pelvic radiograph of exstrophy patient JH (age 49) demonstrating severe hip dysplasia and degenerative changes in both hips.

osteotomy at exstrophy closure reduces symphyseal diastasis, which could lead to a potential reduction in the distance from the body midline to the hip joint, depending on the type of osteotomy. A suprapubic osteotomy[2–3,7–8] would only improve the symphyseal diastasis, but it would not rotate the acetabulae medially as would anterior innominate/posterior innominate osteotomy[4,10]. Osteotomy would change the three dimensional biomechanical aspects of the pelvis as well. In two-dimensional modeling, a reduction in the distance to the body midline from the femoral head would lead to a reduction in joint load and stress and may reduce the risk of early onset of DJD. More advanced biomechanical models and long term follow-up of bladder exstrophy children who have undergone pelvic osteotomy are needed to more fully evaluate the natural history of the hip in bladder exstrophy.

III. SUMMARY

- Force and stress on the hip joint are increased in adult bladder exstrophy patients secondary to:
    1. a significant increase in symphyseal diastasis and thus an increase in the distance from the midline to the hip
    2. significant decrease in distance from the greater trochanter to the center of the hip
    3. significant decrease in CE angle
- The increase joint load and stress can lead to an earlier onset of DJD.
- Closing the pelvic ring at exstrophy closure can help reduce diastasis and the distance from the body midline to the hip joint and change other geometric parameters of the pelvis. This may lead to a reduction in potential joint load and stress and reduce the risk of DJD.

REFERENCES

1. Aadalen, R. J.; O'Phelan, E. H.; Chisholm, T. C.; McParland, F. A. Jr.; and Sweetser, T. H. Jr.: Exstrophy of the bladder: long term results of bilateral posterior iliac osteotomies and two-stage anatomic repair. *Clin. Orthop., 151: 193–200, 1980.*
2. Frey, P.: Bilateral anterior pubic osteotomy in bladder exstrophy closure. *J.Urol.*, 156: 812–815, 1996.
3. Frey, P.; and Cohen, S. J.: Anterior Pelvic Osteotomy. A new operative technique facilitating primary bladder exstrophy closure. *Brit. J. Urol.*, 64: 641–643, 1989.
4. Gearhart, J. P.; Forschner, D. C.; Jeffs, R. D.; Ben-Chaim, J; and Sponseller, P. D.: A combined vertical and horizontal pelvic osteotomy approach for primary and secondary repair of bladder exstrophy. *J. Urol.*, 155: 689–693, 1996.
5. Legal, H.: Introduction to the biomechanics of the hip. In *Congenital Dysplasia and Dislocation of the Hip in Children and Adults*, pp. 26–57. Edited by D. Tonnis. New York, Springer, 1987.
6. O'Phelan, E.H.: Iliac osteotomy in exstrophy of the bladder. *J. Bone and Joint Surg.*, 45-A: 1409–1422, Oct. 1963.
7. Perovic, S.; Brdar, R; and Scepanovic, D.: Bladder exstrophy and anterior pelvic osteotomy. *Brit.J. Urol.*, 70: 678–682, 1992.
8. Schmidt, A. H.; Keenen, T. L.; Tank, E. S.; Bird, C. B.; and Beals, R. K.: Pelvic osteotomy for bladder exstrophy. *J. Pediat. Orthop.*, 13: 214–219, 1993.
9. Sponseller, P. D.; Bison, L. J.; Gearhart, J.P.; Jeffs, R.D.; Magid, D.; and Fishman, E.: The anatomy of the pelvis in the exstrophy complex. *J. Bone and Joint Surg.*, 77-A: 177–189, Feb. 1995.
10. Sponseller, P. D.; Gearhart, J. P.; and Jeffs, R. D.: Anterior innominate osteotomies for failure or late closure of bladder exstrophy. *J. Urol.*, 146: 137–140, 1991.
11. Sutherland, D.; Pike, L.; Kaufman, K.; Mowery, C.; Kaplan, G.; and Romanus, B.: Hip function and gait in patients treated for bladder exstrophy. *J. Pediat. Orthop.*, 14: 709–714, 1994.

4

DETRUSOR MORPHOLOGY AND INNERVATION IN BLADDER EXSTROPHY

A Preliminary Investigation

J. A. Gosling,[1] R. Mathews,[2] and J. P. Gearhart[2]

[1]Department of Anatomy
Faculty of Medicine
The Chinese University of Hong Kong
Hong Kong
[2]Division of Pediatric Urology
James Buchanan Brady Urological Institute
The John Hopkins Medical Institutions
Baltmore, Maryland 21287-2101

INTRODUCTION

The structure of the detrusor in cases of bladder exstrophy has been previously described (Culp, 1964; Engel and Wilkinson, 1970; Rudin et al, 1972) and alterations in the muscularis and an increase in connective tissue have been reported by these workers. It remains uncertain as to whether these structural changes are intrinsic to exstrophy or whether they are secondary to infection or surgical attempts to alleviate the condition. In order to reevaluate these alternatives the present study has employed techniques to examine smooth muscle morphology and its intrinsic autonomic innervation in cases of bladder exstrophy. Particular emphasis has been placed on the distribution of autonomic nerves supplying the detrusor since the nerve supply of bladder smooth muscle in this condition has not been previously described. All specimens were obtained and initially processed by the Division of Pediatric Urology, The Johns Hopkins Medical Institution, Baltimore and then subsequently examined using light microscopy in the Department of Anatomy, Chinese University of Hong Kong, Hong Kong.

The Exstrophy–Epispadias Complex, edited by Gearhart and Mathews.
Kluwer Academic / Plenum Publishers, New York, 1999.

MATERIALS AND METHODS

A total of 42 patients were initially included in the study and of these, tissue samples from 37 cases were fixed in formalin and embedded in paraffin wax. The samples from the 5 remaining cases were rapidily frozen in 2 methylbutane cooled in liquid nitrogen and stored at -75°C in a refrigerator. Paraffin embedded and fresh frozen specimens were then air freighted to Hong Kong for further processing.

Paraffin Embedded Tissues

Serial sections 10μ in thickness were obtained from each sample and every tenth section stained for histology using Massons trichrome technique. Of the 37 cases, biopsy material from only 18 patients (age range, new born–15 years; mean 4 years) possessed sufficient smooth muscle to assess and these were then included for further study. Sections were subsequently processed for histology or for autonomic nerves using immunocytochemical and histochemical methods.

Fresh Frozen Tissue

All samples from the 5 patients (age range 1 year 6 months - 8 years; mean 5 years) were each serially sectioned at 15μ thickness in a cryostat maintained at -25°C internal temperature. Every tenth section was stained using Massons trichrome technique for histology and appropriate intermediate sections processed immunocytochemically or histochemically for autonomic nerves.

Immunocytochemistry and Histochemistry

An indirect fluorescent technique was used to demonstrate protein gene product 9.5 (PGP 9.5) (to study the distribution of autonomic nerves) and dopamine-β-hydroxylase (DβH) (to study the distribution of presumptive noradrenergic nerves). Paraffin sections were placed in xylene and then hydrated using a graded series of alcohol. Both paraffin and cryostat sections were then washed in 3 changes of phosphate buffered saline (PBS) - sucrose containing 0.05% sodium acide, 1% bovine serum albumin and 0.1% Triton and then treated with 1.5% goat serum in PBS for 30 minutes (ABC Kit, Vector Lab, USA) at room temperature.

Sections were then incubated at room temperature for 3 hours with polyclonal antisera to PGP 9.5 (Ultraclone, UK) and DβH (Incstar, USA) raised in rabbits. The optimal dilutions for PGP 9.5 and DβH were 1 in 400 and 1 in 200 respectively. All antisera showed no cross-reactivity to other peptides (maker specifications) and nonspecific interactions were excluded by performing incubations with omission of primary antibodies and by replacing the primary antisera with rabbit serum.

The sites of antigen - antibody reaction were revealed by incubating each section for 1 hour at room temperature with biotinylated antirabbit or antimouse antiserum (BA 1000 and BA 2000 dilution 1:200; Vector Lab, USA) followed by fluorescein isothyocyanate-labelled avidin D (1:300; Vector Lab, USA) for 1 hour. The preparations were washed in 3 changes of PBS, mounted in glycerol/PBS mountant and viewed using a Zeiss microscope equipped with epifluorescence and appropriate excitor and barrier filters. At least 5 sections were examined from each specimen for each antibody studied and the results obtained were found to be consistent for each specimen. Photomicrographs were recorded using Fujichrome (400 ASA) film.

To demonstrate tissue acetylcholinesterase, both paraffin and cryostat sections were processed according to the method described by Gomori (1952) using appropriate inhibitors.

RESULTS

All sections were examined by one investigator unaware of the clinical data of each patient. Subjective assessment of smooth muscle morphology and autonomic innervation was made on arbitary scales and subsequently correlated with the clinical history of each case.

The morphology of the bladder wall in exstrophy could not be correlated with age or clinical presentation of the condition. Thus some biopsy samples possessed histological features indistinguishable from normal detrusor (Figure 1) and the muscle was profusely innervated with PGG 9.5 reactive nerves (Figure 2). In others, some of the smooth muscle bundles were infiltrated with connective tissue (Figure 3) and few nerves were observed (Figure 4) in these regions. However, other bundles within the same bladder biopsy appeared relatively normal (Figure 5) and were associated with numerous autonomic nerves (Figure 6). In those sections in which numerous PGP positive nerves were observed, DβH reactive nerves were associated with blood vessels and few were observed amongst the detrusor smooth muscle bundles (Figure 7). These observations correspond with those obtained in control specimens. apperently

DISCUSSION

The present results have demonstrated the wide spectrum of morphological changes which occur in bladder exstrophy. Nevertheless that some specimens possess an appar-

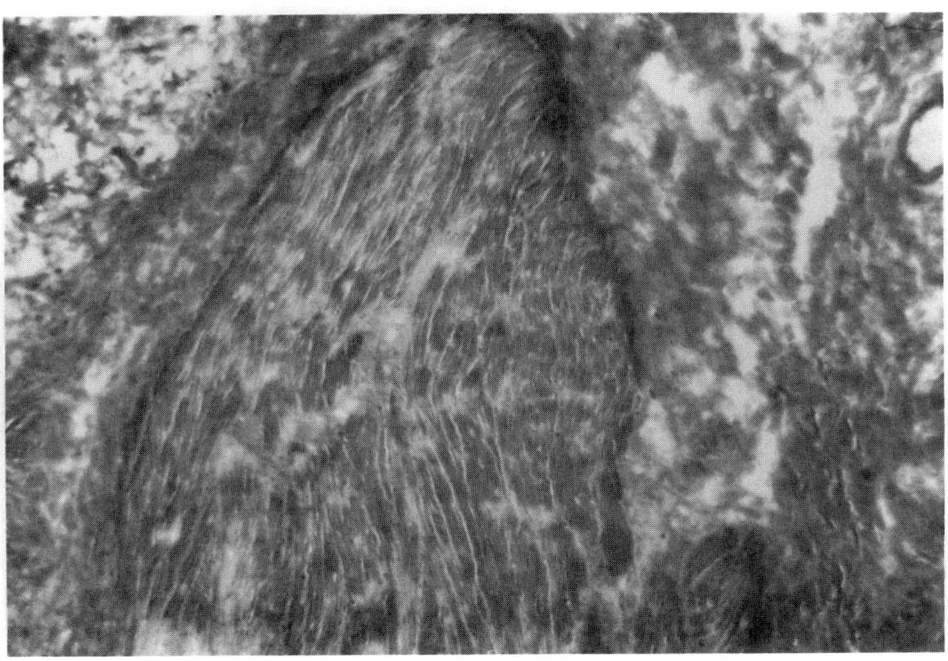

Figure 1. A photomicrophy showing normal detrusor muscle morphology. The bundle consists of closely packed smooth muscle cells. Massons trichrome stain. × 200. (For a color reproduction of this figure, see the insert facing page 20).

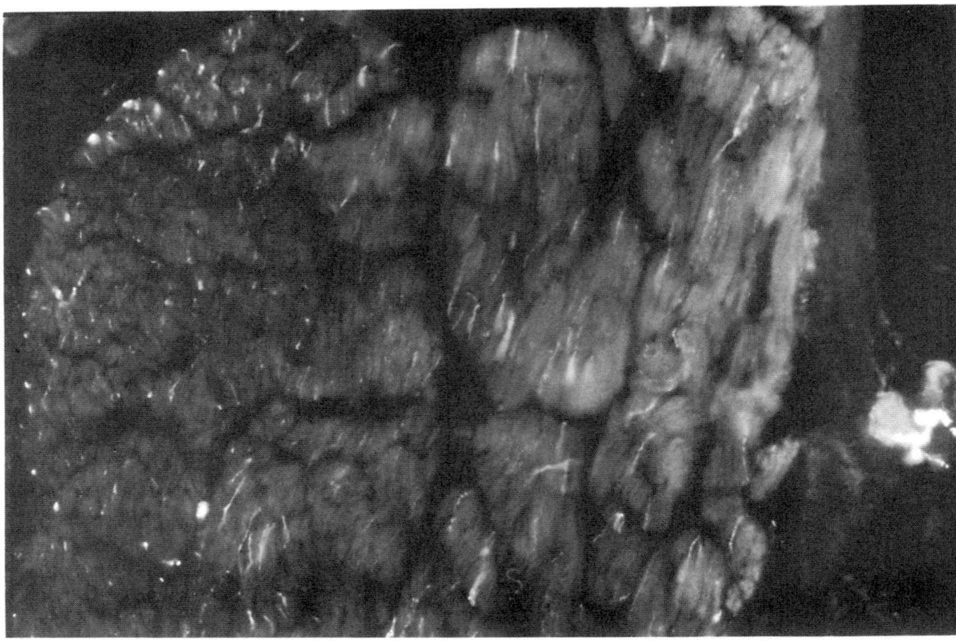

Figure 2. A fluorescence micrograph showing the profuse PGP 9.5 innervation of detrusor smooth muscle in this specimen of exstrophied bladder. × 300. (For a color reproduction of this figure, see the insert facing this page.)

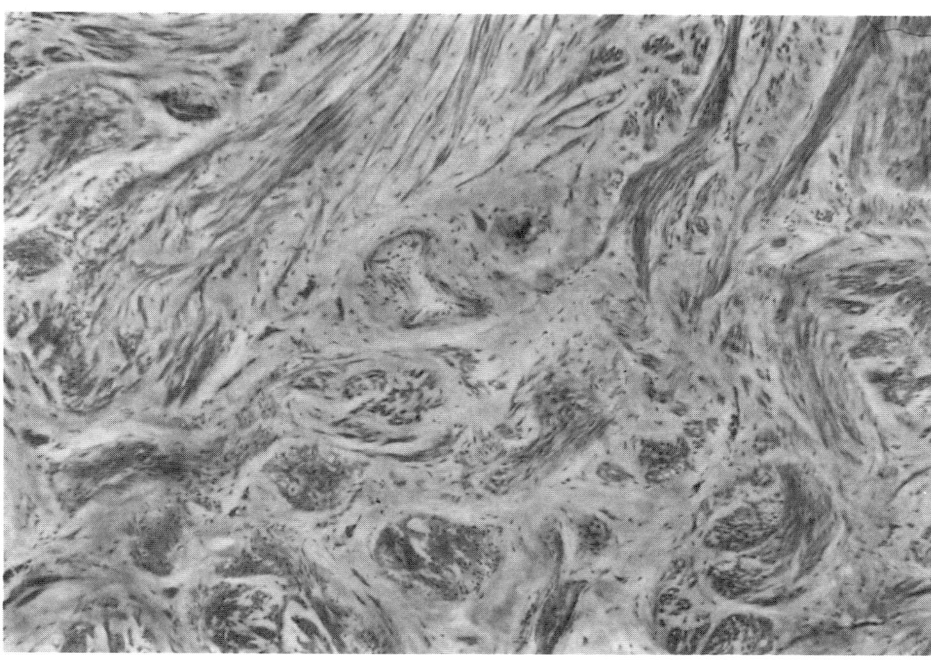

Figure 3. In this sample numerous connective tissue elements pervade the smooth muscle bundles. Massons trichrome strain. × 200. (For a color reproduction of this figure, see the insert facing this page.)

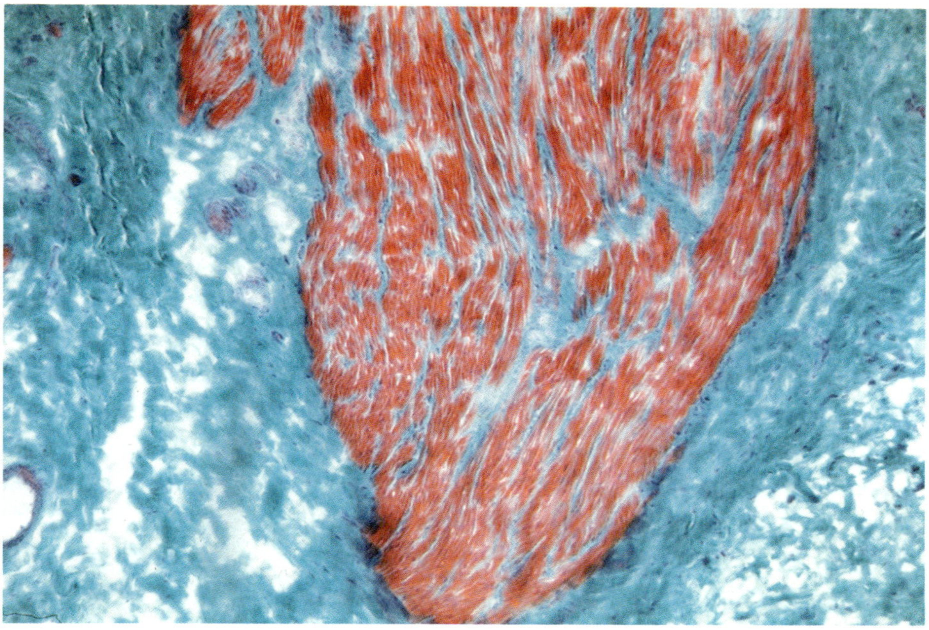

Figure 1.

Figure 2.

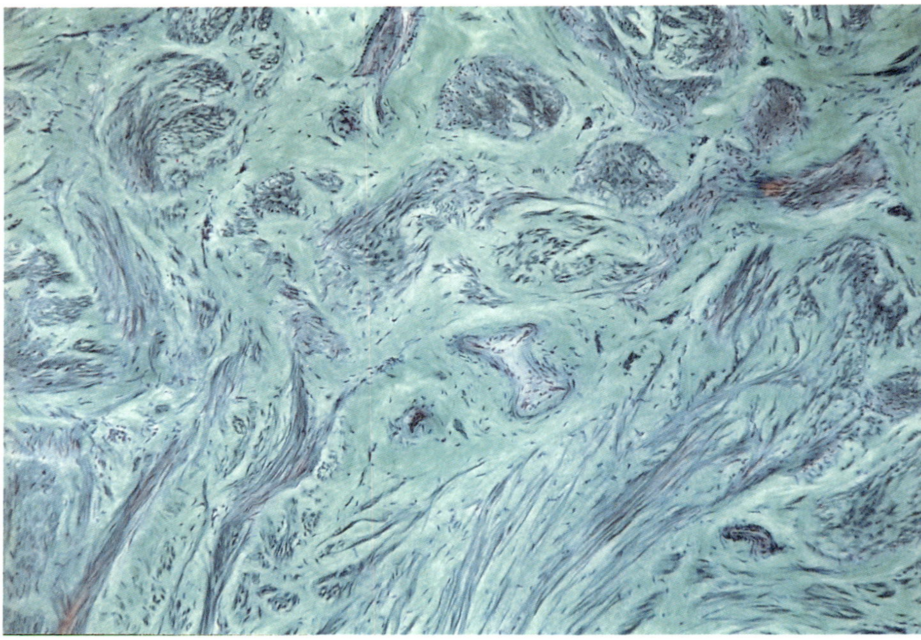

Figure 3.

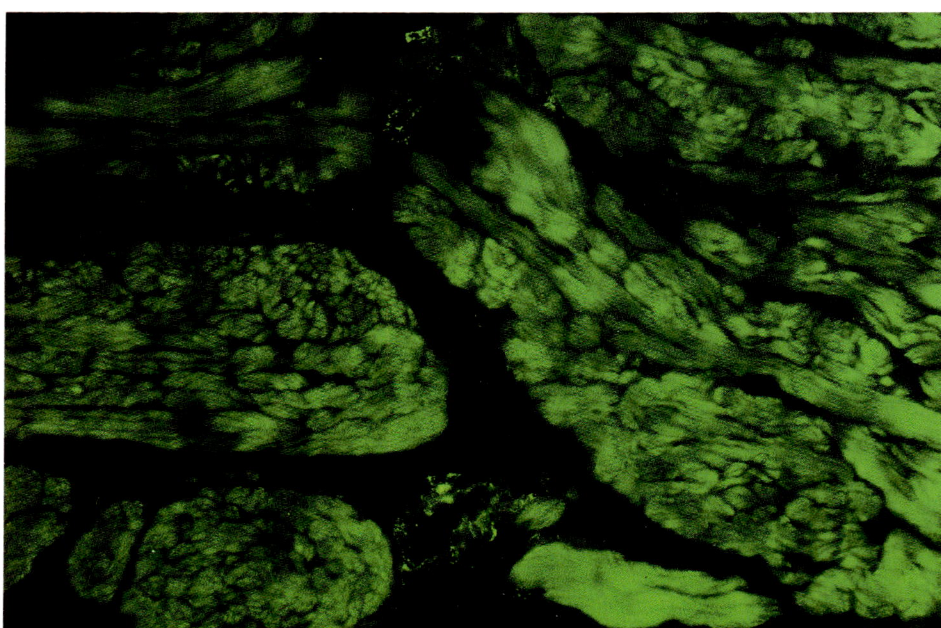

Figure 4.

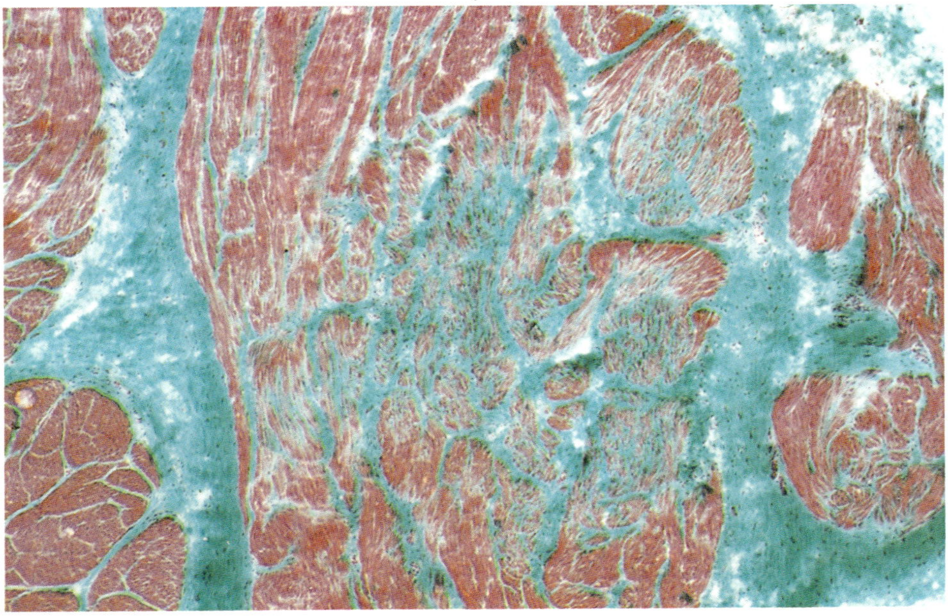

Figure 5.

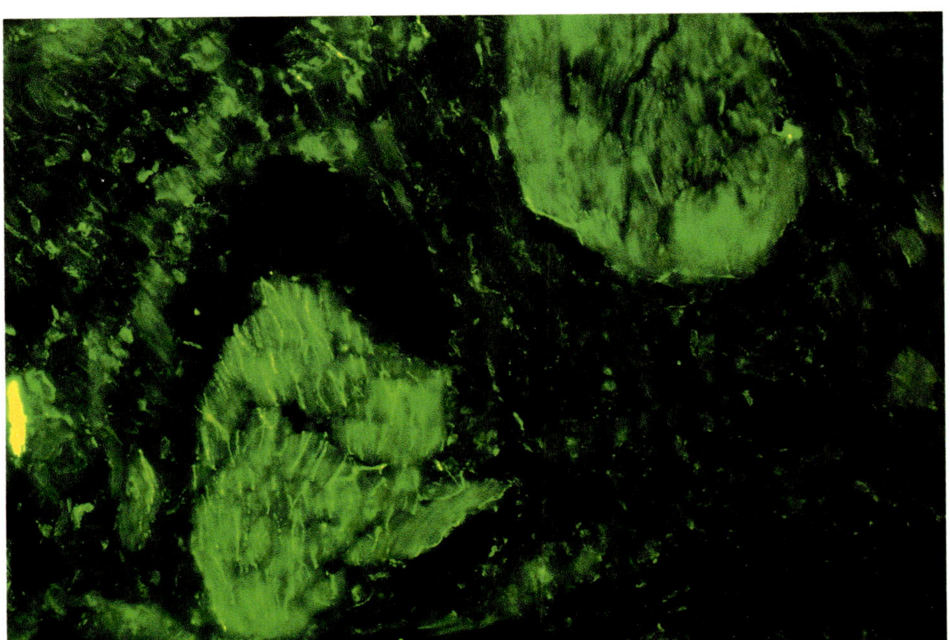

Figure 6.

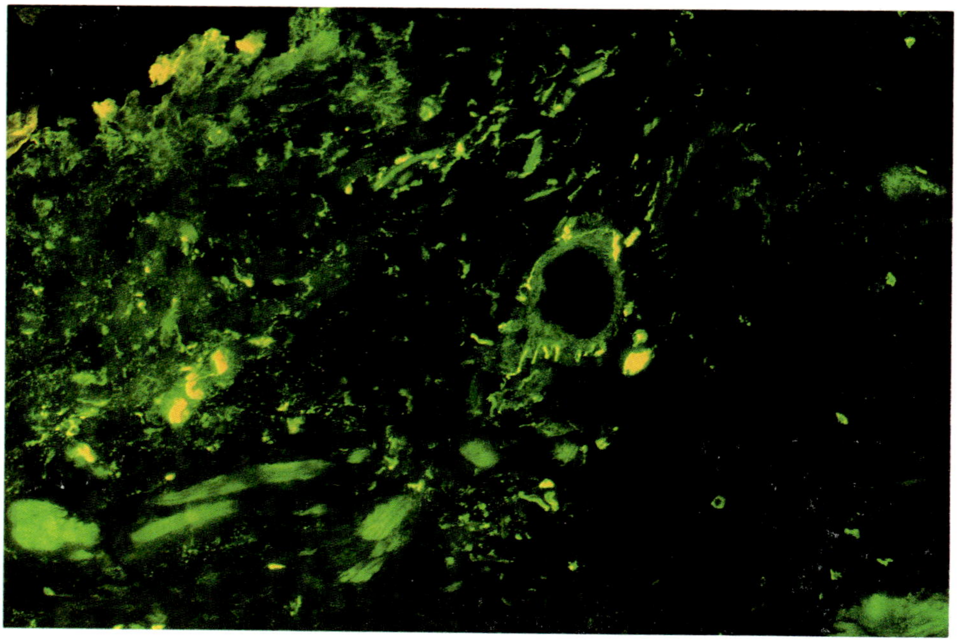

Figure 7.

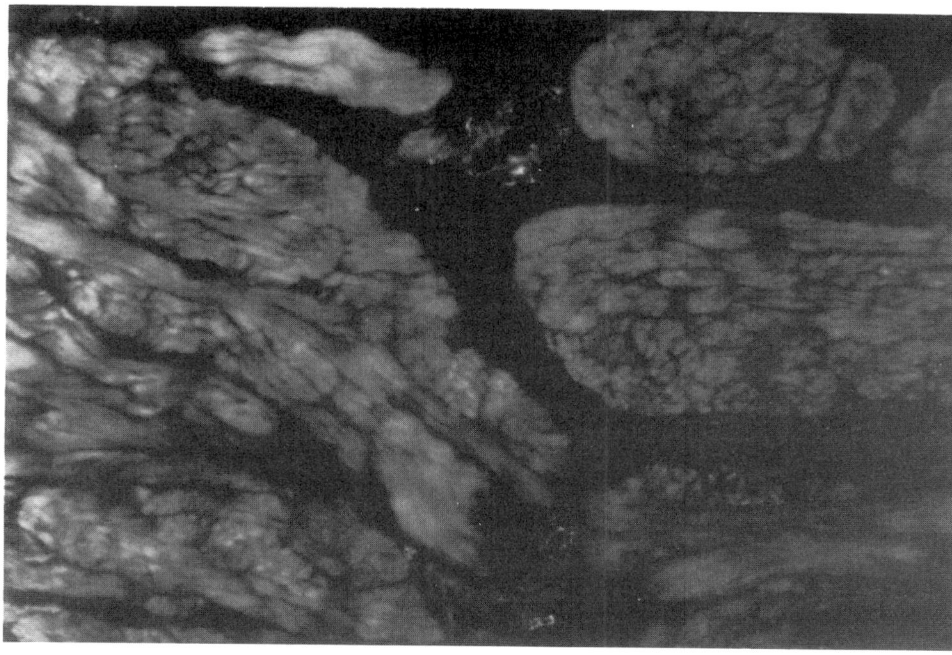

Figure 4. Occasional PGP 9.5 positive nerves are associated with vessels while the detrusor is devoid of similar fibres in this samples of bladder exstrophy. × 200. (For a color reproduction of this figure, see the insert facing this page.)

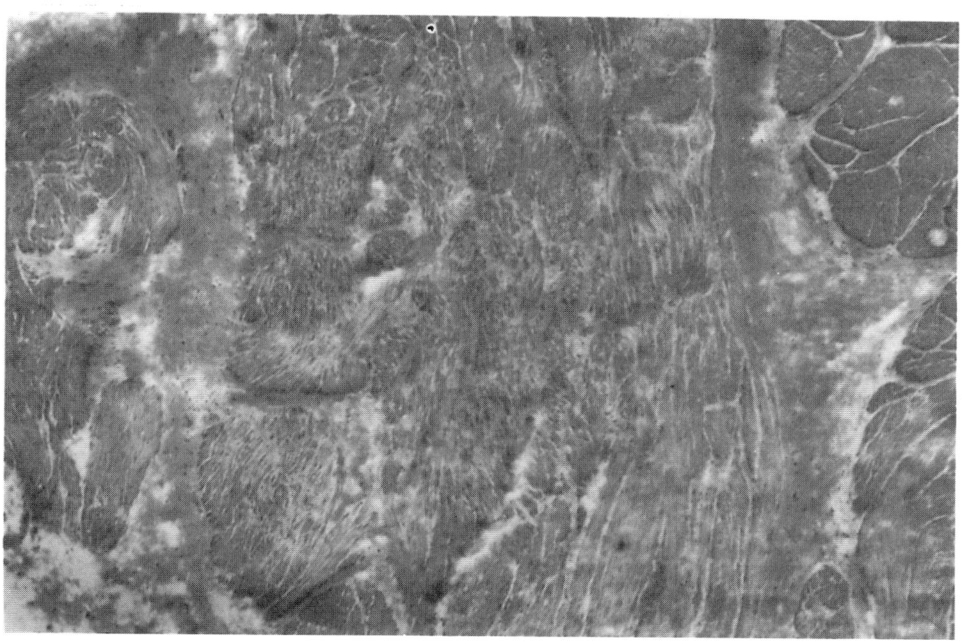

Figure 5. In this specimen some muscle bundles are infiltrated by connective tissue whilst others appear relatively normal. Massons trachrome stain. × 200. (For a color reproduction of this figure, see the insert facing this page.)

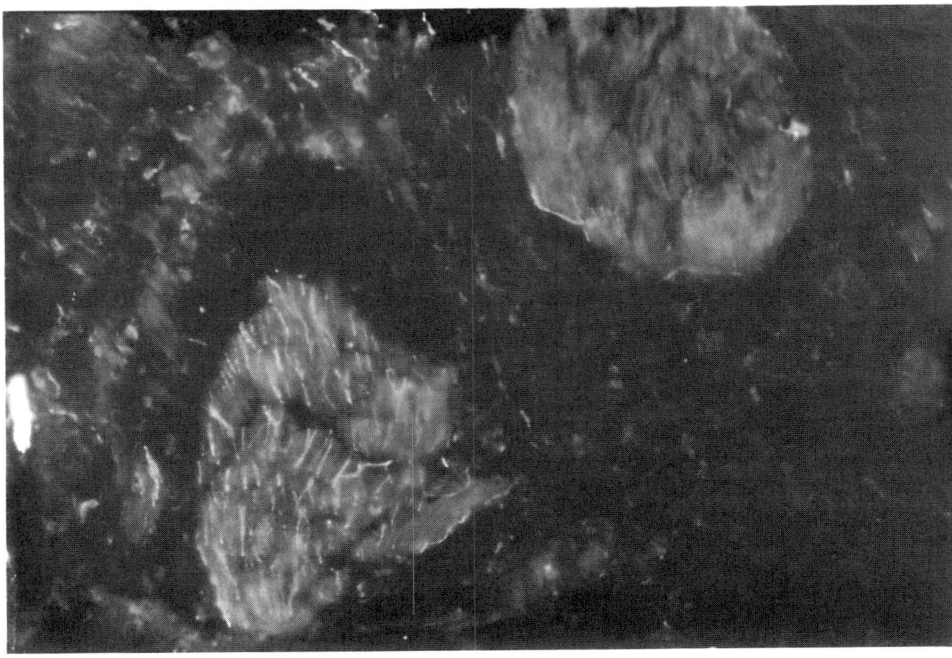

Figure 6. In this photomicrograph one muscle bundle is richly supplied with autonomic nerves while its neighbour is associated with few similar nerves. PGP 9.5 preparation × 300. (For a color reproduction of this figure, see the insert facing page 21.)

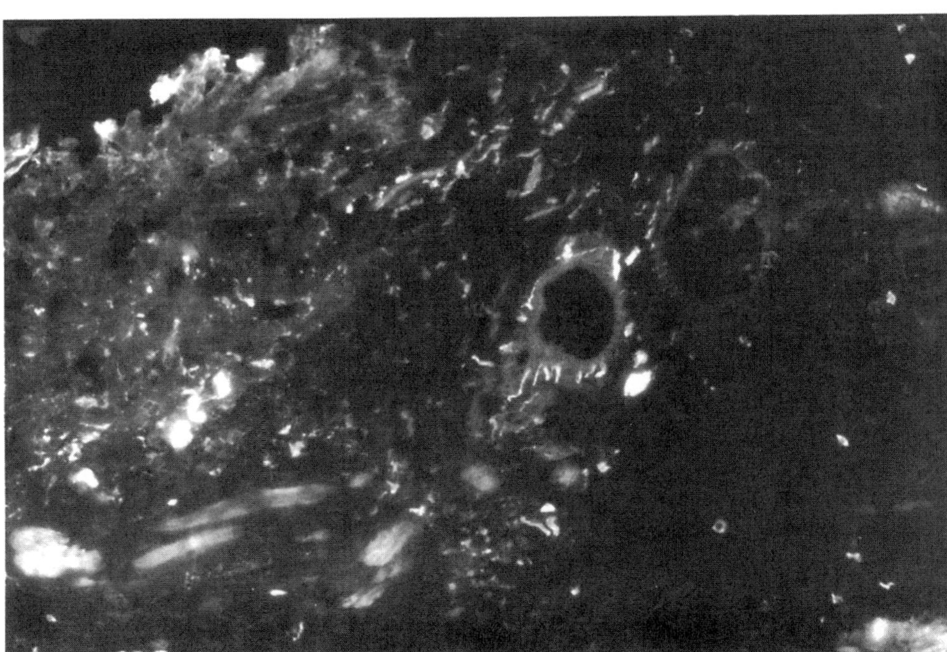

Figure 7. In this DβH preparation most nerves are associated with vessels and are not observed amongst the detrusor smooth muscle cells. × 200. (For a color reproduction of this figure, see the insert facing page 21.)

ently normal detrusor muscle richly innervated by autonomic fibres indicates that the bladder in these cases of exstrophy is capable of adequate function. These findings raise the possibility that an increase in connective tissue and a reduction in autonomic innervation are secondary responses to subsequent events in these particular cases and are not an intrinsic anomaly of the exstrophied bladder. In this regard those specimens showing an increase in connective tissue and a paucity of autonomic nerves are similar to those obtained from bladders subject to outflow obstruction (Gilpin et al, 1985; Gosling et al, 1986). In the present preliminary study the factors responsible for the increase in connective tissue and apparent loss of autonomic innervation cannot be determined. However the severity of these changes in some of the cases indicate that neural control and contractile capability of the detrusor is likely to be very inadequate. Should further surgery be attempted in these patients it seems likely that the outcome will be seriously jeopardized by an already compromised detrusor. In this context the present results indicate the need for histological evaluation of all cases in which reconstructive surgery is to be attempted. In conclusion the present study has indicated the need for a morphological evaluation of the exstrophied bladder prior to surgery in order to assess the status of the detrusor in each case. By so doing an indication of the functional capability of the reconstructed bladder will be obtained. A further indepth study is required in order to determine those factors which contribute to the increase in connective tissue and the reduction in autonomic nerves associated with some cases of bladder exstrophy.

REFERENCES

Culp, D.A. The histology of exstrophied bladder. J. Urol., *91*, 538–548, 1964.

Engel, R.M. and Wilkinson, H.A. Bladder extrophy. J. Urol., *104*, 699–704, 1970.

Gilpin, S.A., Gosling, J.A. and Barnard, R.J. Morphological and morphometric studies of the human obstructed, trabeculated urinary bladder. Brit. J. Urol., *57*, 525–529, 1985.

Gomori, G. Microscopic histochemistry: principles and practice. University of Chicago Press, Chicago, 1952.

Gosling, J.A., Gilpin, S.A., Dixon, J.S. and Gilpin, C.J. Decrease in the autonomic innervation of human detrusor muscle in outflow obstruction. J. Urol., *136*, 501–504, 1986.

Rudin, L., Tannenbaum, M and La Himer, J.K. Histologic analysis of the exstrophied bladder after anatomical closure. J. Urol., *108*, 802–807, 1972.

5

THE RESPONSE OF THE BLADDER SMOOTH MUSCLE TO INCREASED BLADDER NECK RESISTANCE

Edward J. Macarak, Stephen Zderic, Howard M. C. Snyder, Douglas Canning, and Pamela S. Howard

University of Pennsylvania, and
The Division of Pediatric Urology
Children's Hospital of Philadelphia, Pennsylvania

The following series of studies are the result of a long association with the Division of Pediatric Urology at the Children's Hospital of Philadelphia. This division, since its inception, was headed by Dr. John W. Duckett. John's innovative skills as a surgeon and his quest to improve the outcome of the children in his care led him to initiate a fellowship program to train the next generation of pediatric urologists. It has been a privilege to be a part of this program and to know and work with John Duckett. This chapter is respectfully dedicated to his memory and to his wonderful legacy to Pediatric Urology.

Increased bladder outlet resistance presents a chronic problem with respect to the normal mechanics of bladder function. Normally the bladder fills with little or no increase in infravesical pressure and then expels the urine at an appropriate time. Afferent signals, likely from stretch receptors associated with nerve endings in the bladder wall, trigger the micturition reflex. However, if there is increased resistance at the bladder neck, the bladder must correspondingly increase the pressure within the bladder to open the urethra and expel the urine. If the bladder is forced to operate under these conditions for extended periods of time, compensatory changes occur with respect to physiologic function and matrix composition.[1,2,3,4,5] The molecular mechanisms responsible for these changes is currently unknown.

Outflow obstruction can occur as a result of a physical obstruction at the bladder neck or proximal urethra. In some instances, these obstructions can be of an acute and congenital nature, e.g., posterior urethral valves. Obstruction can also occur chronically. Most males over the age of 40 experience mild degrees as obstruction as a result of benign prostatic hyperplasia. Other forms of outlet obstruction occur which can best be described as functional obstructions, e.g., detrusor-sphincter dyssynergia. In this instance, the obstruction results not from a physical obstruction but rather from a functional inability to

The Exstrophy–Epispadias Complex, edited by Gearhart and Mathews.
Kluwer Academic / Plenum Publishers, New York, 1999.

coordinate the actions of the sphincter and detrusor muscle. In effect, it is a form of obstruction similar to anatomic ones because in all cases, the bladder is forced to work against an increased outlet resistance. Thus, while the etiology of these conditions may be different, the end results on bladder structure and function are quite similar.

A major clinical symptom associated with outlet obstruction of any type is some form of bladder instability.[6] Bladder instability manifests itself in the form of involuntary contractions or other aspects of increased bladder sensitivity that alter the normal cycling regimen of the bladder. With increasing instability, the bladder wall is subjected to a different regimen of cycling, i.e., it is forced to spend greater portions of the filling cycle at higher pressures. This integral of time and pressure is an important measure of normal cycling. In addition to sustaining higher infravesical pressures, the wall may be experiencing involuntary contractions subjecting the constituent cells to a variety of abnormal mechanical signals. Whatever the underlying cause, as a result of obstruction, the cells within the bladder wall begin receiving abnormal "cues" and respond to them. Stimulation of cells can occur by autocrine or paracrine secretion of cytokines, neurotransmitters, hormones and/or other "effector" molecules.[7] In addition, increased or prolonged mechanical "stretch" of the bladder wall can alter channels within the cell membranes of nerve or other bladder wall cells.[8] Ultimately these signals, regardless of their origin or associated pathology, alter the phenotype of bladder wall cells. It is likely that the altered phenotypic expression of bladder wall cells leads to some of the physical and compositional changes that occur in response to outlet obstruction.

Initially these changes are largely reversible.[9] If the obstruction is relieved before the bladder undergoes "decompensation", both structural and functional changes almost completely disappear.[10] Thus there appears to be a "window" during which intervention to either surgically or perhaps pharmacologically remove the obstruction is beneficial to the long term outcome. Animal studies from a number of laboratories have demonstrated that the initial response of the bladder wall to obstruction can best be described as a compensatory response. That is, the bladder musculature responds to the increased work load caused by the obstruction by increasing muscle mass. Biologically, the detrusor is being forced to increase its contractile force to overcome the increased resistance which is accomplished by synthesizing more contractile proteins. There is an almost immediate increase in muscle mass that occurs as a result of muscle hypertrophy and/or hyperplasia. Biochemically, the level of actin and myosin increases as well as other associated contractile proteins that alter the contractility of the detrusor.

In addition to changes in contractile proteins, alteration in the connective tissue composition of the bladder wall also occurs. While these changes are substantial, they are not as great as those in contractile protein. Measurements of collagen usually show a significant absolute increase in collagen; however, relative to the more substantial increase in overall mass resulting from detrusor hypertrophy, such changes are usually not significant.

Changes in collagen manifest themselves in two ways: 1) a change in the distribution and 2) a change in the amount. To evaluate these changes, our strategy has been to employ immunohistochemical, biochemical and molecular biological approaches. Alteration in collagen architecture is best studied using immunohistochemistry with collagen type-specific antibodies. Previous studies from our laboratory as well as those from other laboratories have clearly demonstrated that changes in collagen composition accompany various types of outlet obstruction. Ewalt et al.[11] showed that neurogenic bladders show altered distribution of type III collagen. In Figure 1, a section from a normal bladder has been stained with an antibody to type III collagen. The section shows several muscle bundles with localization of the type III collagen largely confined to the perimysial region (arrow) surounding the detrusor muscle bundles (MB). Little or no type III collagen is found

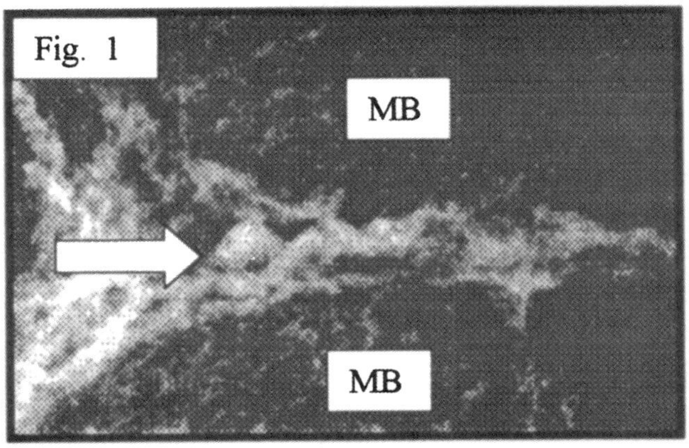

Figure 1.

within muscle bundles. In contrast, type III collagen redistributes from a perimysial to an endomysial location (Small arrows) in a subset of patients with neurogenic bladders (Figure 2). These changes appears to be specific for type III collagen since type I collagen shows no similar change in distribution. Thus sections through noncompliant bladders characteristically show some but not all muscle bundles with type III collagen located intrafasiculcularly.

Thus immunohistochemistry can demonstrate changes in a particular collagen type and location in selected patients. While such information is of obvious value, it is difficult to characterize these changes without some means to objectively evaluate them. To obtain such information, it is necessary to biochemically evaluate these same changes by applying techniques that lend themselves to their quantification. To evaluate and quantify changes in collagen, tissue samples are solubilized using cyanogen bromide. This reagent cleaves proteins containing methionine making the resulting peptides soluble. When bladder tissue is treated with this reagent, approximately 95% of the tissue weight is solubilized. The small residue that remains undissolved is largely elastin which does not contain the amino acid methionine and is therefore resistant to cyanogen bromide digestion. This approach was used to study a group of patients with non-compliant bladders.[12]

In this study, tissue samples from pediatric patients with either functional (largely neurogenic bladders) or anatomic (largely posterior urethral valves), were biochemically

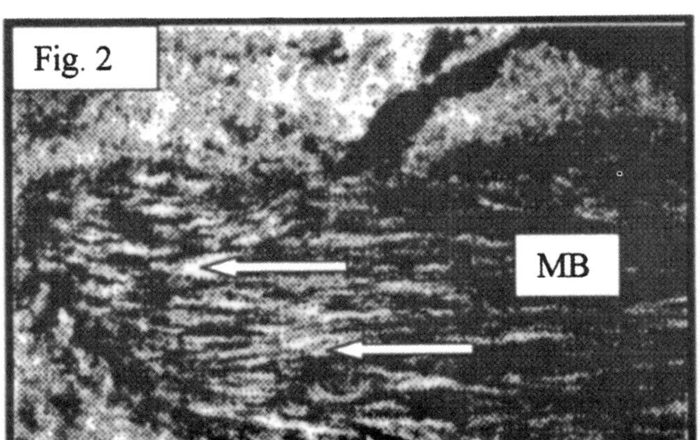

Figure 2.

Table 1. Collagen quantification

	Collagen Mg/gm dry wt	% Type I	% Type III	Type I Mg/gm dry wt	Type III Mg/gm dry wt
Normal	263 ± 60	76.2 ± 3	23.8 ± 3	200.7 ± 49	62.6 ± 15
Non- Compliant	289 ± 52	67.2 ± 11	32.8 ± 11	193.6 ± 53	93.1 ± 36
Change	9.8%	12%	37.8%	-3.67	48.7%

analyzed to evaluate changes in matrix composition. Results show an overall increase in hydroxyproline content in dysfunctional bladders; although, not surprisingly, the change is not significant when normalized to overall bladder weight. However, analysis of the absolute increase in collagen, that is the increase in collagen mass relative to the basal mass, is significant. Regardless of whether either neurogenic and nonneurogenic bladders are analyzed separately or combined as a group, Type III collagen increases by approximately 60% (Table 1). Thus the biochemical and immunohistochemical results are in agreement. These data demonstrate that both functional and anatomic obstruction can lead to an alteration in not only the location but also the quantity of type III collagen.

Since the major fibrillar collagens found in the bladder wall are types I and III collagen, a consequence of these changes is an alteration in the type III: type: I collagen ratio (Table II) . In normal bladders, the collagen composition is approximately 25% type III collagen and 75% Type I collagen (Type III:I ratio = 1:3) In bladders with progressive fibrotic changes, this ratio shifts to approximately 33% type I and 66% Type III collagen (Type III:I ratio = 1:2). While little is currently known about the functional consequences of such changes, there are several possibilities that relate to the mechanical properties of the tissue.

To begin to establish information on a potential mechanism for this fibrotic response, mRNA was extracted from the same bladder tissue analyzed above. These mRNAs were then probed with DNA fragments specific for type I and type III collagen. Signals for individual probes were normalized for probe size and signal intensity. The results of these studies, are shown in Table 3. The ratio of steady state levels of the $\alpha1$[III]: $\alpha1$[I] mRNAs show a significant increase for neurogenic but not non-neurogenic bladder tissue. Thus, for neurogenic but not non-neurogenic bladders, an increase in mRNA levels yields a corresponding increase in protein levels in a coordinate fashion. In nonneurogenic bladders, there is an increase in the $\alpha1$[III]: $\alpha1$[I] ratio, but this change is not significant. In summary, immunohistochemical and biochemical analyses of neurogenic and non-neurogenic bladder tissue demonstrate changes type III collagen location and quantity. These changes, at least in neurogenic bladders, are transcriptionally regulated since steady state mRNA levels also show a corresponding rise. For non-neurogenic bladders, regulation appears more complex since there was not a direct correlation between mRNA and protein levels.

The observation that bladder wall fibrosis is characterized by infiltration of some detrusor muscle bundles with increased quantities of type III collagen has potential functional significance. If alteration in type III collagen occurs and is associated with changes

Table 2. Type III: Type I ratio

Noncompliant	Control
32.8% : 67.2%	23.8% : 76.2%

Table 3. Collagen mRNA analysis

	Control (n=10)	Non-neurogenic (n=9)	Neurogenic (n=25)
α1[III]: α1[I] ratio	2.99 ± 0.12	0.34 ± 0.14	0.454 ± 0.23*

* p = 0.049

in the functional compliance properties of the bladder wall, what role do these fibers play in the architecture of the bladder wall and in tension transfer?

To begin to answer this question, a series of observations were made on partially filled bladders. For these experiments, bovine bladders were filled to full or partial capacity and then frozen and sectioned through the bladder dome. Sections were stained with antibodies to type III collagen and the position and form of type III collagen fibers noted. While the fibers are clearly visualized, it is difficult to evaluate their internal conformation (data not shown). To further refine these observations, additional observations were performed using a laser confocal microscope.

Using the confocal microscope, the internal conformation of the fibers can clearly be visualized[13]. As shown in Figure 3, fibers containing type III collagen exhibit a characteristic "coiled configuration." When tissue from bladders at different levels of filling were examined, it was observed that the tortuosity of the coils was varied depending upon the state of filling. For example, in empty bladders, coils of type III collagen were compact and showed maximum tortuosity (Figure 4). In partially filled bladders, type III collagen fibers were less tortous and had an expanded coil structure. As shown in Figure 5, these expanded coils were slightly thinner than the coils observed in the empty bladder. Finally, when type III fibers in full bladders were examined, the collagen coils showed the least tortuosity and had the most expanded coil structure. Thus, the period of the collagen coiling (the distance from the "peak" of one coil to the "peak" of the next coil) is smallest in the undistended bladder (greatest tortuosity) and is the largest in the bladders filled to capacity (least tortuosity).

SUMMARY

Observations from our laboratory over the past seven years suggest that various forms of outlet obstruction result in a form of fibrosis within the detrusor muscle layer. The fibrosis is the result of an increase in type III collagen that accumulates within rather than around some muscle bundles. This increased and altered expression of type III collagen appears to functionally disrupt the ability of the bladder to function as a low compli-

Figure 3.

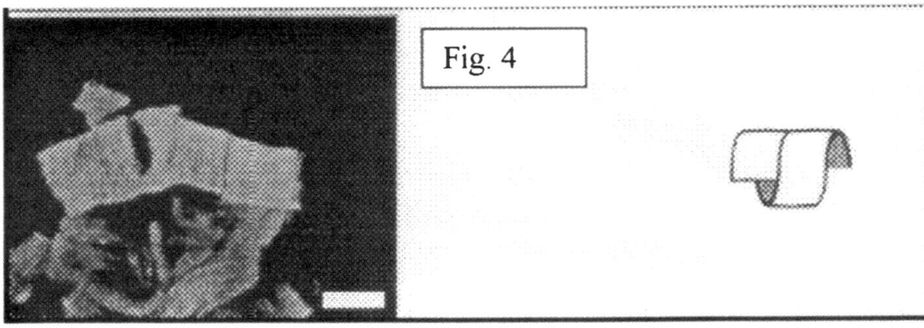

Figure 4.

ance structure although the exact mechanism by which increased type III may alter compliance is unknown. One possibility is that increased synthesis of type III collagen by bladder wall cells may alter the mechanical properties of the fibers that transfer tension within the wall. It has previously been demonstrated that collagen fiber diameter can be altered by changing the proportion of type III and type I collagen or by type V and type I.[14] Thus, if one assumes that mechanical strength or some other physical property is related to fibril diameter, it is obvious, based upon our biochemical data, that such changes could and do occur as a result of obstruction.

This, however, raises the question of how volume accommodation actually occurs in the bladder. If type III collagen is a tension bearing element within the bladder wall, then, as the bladder fills, the collagen fibers should stretch or change their shape to accommodate the increased tension that develops. If the tendon is used as an analogy, it can be appreciated that as the tendon is "loaded", its length stretches to a new length that is approximately 4–6% longer. The use of the word stretch is actually incorrect here since the collagen fibers do not actually stretch but more correctly "straighten out" as they begin to carry tension. In the unloaded state, they have a very slightly coiled conformation that, upon loading, is lost. The increased length thus results from a loss of the slight waviness of tendon collagen fibers rather than their stretching.

In contrast to the tendon, the bladder wall undergoes a much greater expansion than 4–6%. How then does the bladder wall and the collagen fibers that comprise its structural architecture actually sustain such large swings in distension? Based upon our observations with the confocal microscope, we suggest that bladder wall tension increases during bladder filling. In fact, this tension may be a necessary factor in the normal development of the bladder.[15] This tension is dissipated within the structure of the bladder wall by expansion of the coil structure of type III collagen. Fibers in an empty and undistended bladder are highly tortous with each individual coil directly adjacent to another. In contrast, a full bladder has collagen fibers whose coil structure is extended or expanded and is much less tortuous that in the empty bladder. Such coils can easily accommodate large tensional

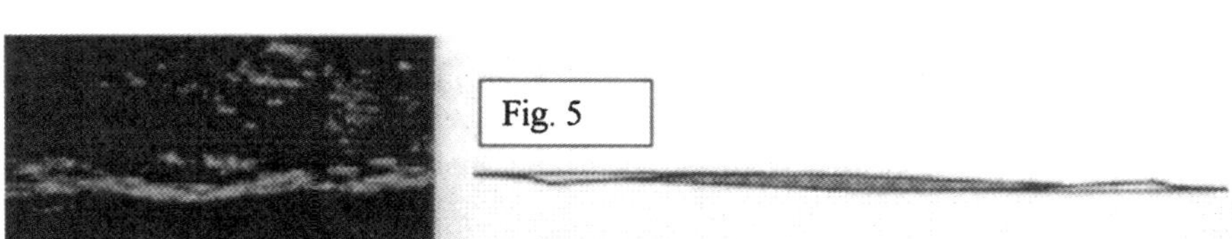

Figure 5.

changes with little or no danger of its fibers being extended to such a great extent that they "plastically or irreversibly" deform and fracture. The role of type I collagen must also be considered. Although we have focused on the role of Type III collagen in this study, Type I collagen is the major structural collagen found in the bladder.

Recent evidence suggests that the presence of Type III is critical for Type I fibrillogenesis in the cardiovascular systems and other organs as well[16]. Thus there is a need for more information about the interaction between these two important structural proteins.

While this model is speculative, it is attractive in that it can account for the physiologic properties of the bladder. It also implies that connections must exist between the tension generating elements, i.e., the smooth muscle cells, and the other components of the bladder. In bladders that become noncompliant, it is likely that there is some interference with the ability of the collagen fibers to elastically and reversibly alter their tortuosity. This, predictably, would reduce total bladder capacity. Further studies will be required to establish the relationship between compliance changes and the passive mechanical elements of the bladder wall that comprise its structural protein matrix.

REFERENCES

1. Cass, S.C. and Hinman, F. 1976. Constant urethral flow in female dog. II. Effect of constriction of vesical neck and external meatus. J Urol., *99*: 447–454.
2. Uvelius, B., Peterson, L and Mattiasson, A. 1984. Smooth muscle hypertrophy and hyperplasia in the rat detrusor after short-time infravesical outflow obstruction. J. Urol., *131*: 173–176.
3. Levin, RM, High, H and Wein, AJ. 1984. The effect of short-term obstruction on urinary bladder function in the rabbit. J. Urol., *132*: 789–791.
4. Kim, KM, Kogan, BA, Massad, CA and Hyang, Y-C. 1991. Collagen and elastin in the obstructed fetal bladder. J Urol., *146*:528–531.
5. Nielsen, KK, Andersen, CB, Petersen, LK, Oxlund, H and Nordling, J. 1995. Morphological, stereological and biochemical analysis of the mini-pig urinary bladder after chronic outflow obstruction and after recovery from obstruction. Neurourology and urodynamics, *14*: 269–284.
6. Elbadawi, A. 1995. Pathology and pathophysiology of detrusor in incontinence., Urologic Clinics of North American, *22*: 499–512.
7. Levin, RL. 1996. Developmental Aspects of urinary bladder physiology and pharmacology. In: Dialogues in Pediatric Urology, ed, Richard Ehrlich, William J. Miller Assoc., Inc, Pearl River, New York, pp. 2–3.
8. Sachs, F. 1997. Mechanical transduction by ion channels: how forces reach the channel. Society of General Physiologists Series. *52*:209–18.
9. Gabella, G and Uvelius, B. 1994. Reversal of muscle hypertrophy in the rat urinary bladder after removal of urethral obstruction. Cell Tissue Res., *277*:333–339.
10. Chen, M-W, Levin, RM and Buttyan, R. 1995. Peptide growth factors in normal and hypertrophied bladder. World J Urol., *13*: 344–348.
11. Ewalt DH. Howard PS. Blyth B. Snyder HM 3d. Duckett JW. Levin RM. Macarak EJ. 1992. Is lamina propria matrix responsible for normal bladder compliance?. J Urol., *148*:544–549.
12. Deveaud, CM, Macarak, EJ, Kucich, U, Ewalt, DH, Abrams, WR and Howard, PS. 1998. Molecular analysis of collagens in bladder fibrosis., J. Urol., (In Press).
13. Chang, SL, Howard, PS and Macarak, EJ. 1998. Role of type III collagen in bladder filling. Neurology and urodynamics, *17* (In Press).
14. Keene DR, Sakai LY, Bachinger HP and Burgeson RE. 1987. Type III collagen can be present on banded collagen fibrils regardless of fibril diameter. Journal of Cell Biology. *105*: 2393–2402.
15. Close CE. Carr MC. Burns MW. Mitchell ME. 1997. Lower urinary tract changes after early valve ablation in neonates and infants: is early diversion warranted? J.Urol. 157:984–988.
16. Liw, X, Wu, H, Byrne, M, Krane, S and Jaenisch, R. 1997. Proc. Nat. Acad. Sci. USA, *94*: 1852–1856

6

MUSCARINIC CHOLINERGIC RECEPTORS IN BLADDER EXSTROPHY

Pharmacological Observations Supporting Functional Closure and Subsequent Bladder Neck Reconstruction

Ellen Shapiro,[1] Robert D. Jeffs,[2] John P. Gearhart,[2] and Herbert Lepor[1]

[1]New York University School of Medicine
New York, New York, and
[2]The Department of Urology
The Johns Hopkins University School of Medicine
Baltimore, Maryland

ABSTRACT

The surgical management of classical bladder exstrophy should be influenced by the inherent detrusor function of the exstrophied bladder. Up until the last decade, the biochemical and neurophysiological properties of the exstrophied bladder had never been studied. Our study using radioligand receptor binding techniques was the first investigation to compare the density and equilibrium dissociation constant of muscarinic cholinergic receptors in control and exstrophy bladders. The density of muscarinic cholinergic receptors in the control and exstrophy groups was 1.97 ± 0.29 and 1.44 ± 0.21 fmol. per μg. DNA (mean ± S.E.M.), respectively. The dissociation constant of the control and exstrophy groups was 0.15 ± 0.02 and 0.14 ± 0.02 nM. (mean ± S.E.M.), respectively. The data show that the muscarinic receptor density and binding affinity in control and exstrophy bladders are similar. Therefore, the neurophysiological composition of the exstrophied bladder is not significantly altered during the anomalous development. Quantitative morphometric studies suggest that the closed bladder sxstrophy has a cellular composition that is similar to control bladders. Our findings together with these morphometric studies suggest that patients with bladder exstrophy should have normal detrusor contractility if successful bladder closure and subsequent bladder neck reconstruction are ultimately achieved.

The Exstrophy–Epispadias Complex, edited by Gearhart and Mathews.
Kluwer Academic / Plenum Publishers, New York, 1999.

INTRODUCTION

The primary objectives of the surgical management of bladder exstrophy are to provide a secure abdominal wall closure, to obtain urinary continence with preservation of renal function and to reconstruct a cosmetically acceptable penis in the male subject. Abdominal wall closure and urinary continence can be achieved following functional bladder closure or urinary diversion. Since both approaches for managing exstrophy have been fraught with complications,[1,2] a consensus for the surgical management of bladder exstrophy had not been reached until the mid-1980's.

Marshall and Muecke reviewed 329 functional bladder closures reported in the literature between 1906 and 1966, and determined that urinary continence with preservation of renal function was achieved in only 5 per cent of the case.[2] Nisonson and Lattimer demonstrated normal detrusor function in only 22 percent of closed exstrophy bladdders.[3] These observations questioned the functional capability of the exstrophied bladder and the rationale for functional closure. Over the past decade, Jeffs and associates have reported the achievement of urinary continence in approximately 80 per cent of consecutive patients who had undergone bladder closure for exstrophy.[4–5] Urodynamic assessment of continent individuals following successful bladder closure and bladder neck plasty identified normal reflexive bladders and normal anal plug electromyograms in 70 and 90 per cent, respectively,[6] of continent closed exstrophies. The high success rates following staged reconstruction implies that the exstrophied bladder is capable of normal function. [4–6]

Normal bladder function is controlled by the autonomic nervous system. Neurotransmitter substances are released at postganglionic synapses and these substances bind to specific neurotransmitter receptors associated with the detrusor muscle. Detrusor function is mediated primarily by muscarinic cholinergic receptors.[7] To evaluate the neuropharmacological properties of exstrophy bladders in 1984, we compared the muscarinic cholinergic receptor density of exstrophy bladders at the time of bladder neck plasty vs. age-matched controls undergoing ureteroneocystostomy.[8] Our detailed methodology for characterizing muscarinic cholinergic receptor density in genitourinary tissues was previously reported. The objective of this investigation of the neurophysiological properties of exstrophied bladders was to provide insight into the functional potential of these bladders after closure.

The materials and methods section and the results of this study are produced from the original manuscript. No other investigators have characterized the neuropharmacology or neurophysiology of the exstrophied bladder. This rereview of this topic provides the opportunity to interpret our findings in light of morphometric observations related to the exstrophied bladder.

MATERIALS AND METHODS

Control bladder tissues were obtained from 6 patients with ureterovesical reflux at the time of ureteroneocystostomy. Normal bladder function and appearance on voiding cystourethrogram were demonstrated in these patients preoperatively. Bladder specimens were obtained from 6 patients with classical bladder exstrophy at the time of the Young-Dees-Leadbetter bladder neck plasty. All children had uninfected urine preoperatively. The mean ages of the control and exstrophy patients were similar (see Table 1). All tissues were weighed immediately and stored individually in a liquid nitrogen freezer.

Table 1. Muscarinic cholinergic receptors in control and exstrophied bladders

Pt. No.—Sex—Age (yrs.)	Equilibrium Dislocation Constant (nM.)	Specific [^{3}H] N-Methylscopolamine Receptor Binding (fmol./μg. DNA)
*Control**		
1—M—6	0.11	
2—F—6	0.19	1.04
3—M—2	0.11	1.76
4—F—6	0.19	3.06
5—F—6	0.15	2.03
6—F—4	0.15	2.36
Mean ± standard error	0.15 ± 0.02	1.97 ± 0.28
Exstrophy†		
1—F—3	0.24	2.45
2—M—4	0.09	1.01
3—M—5	0.12	1.14
4—M—5	0.11	1.48
5—M—4	0.12	1.42
6—M—3	0.14	1.16
Mean ± standard error	0.14 ± 0.02	1.44 ± 0.21

*Mean age 5.0 ± 0.7 years.
†Mean age 4.0 ± 0.4 years.

RADIOLIGAND RECEPTOR BINDING METHODS

Tissue Preparation

Bladder specimens were immersed into liquid nitrogen and pulverized with a Thermovac tissue pulverizer (Redi-Industries Corp., Copiaque, NY). For each saturation experiment 300 to 500 mg.of the pulverized bladders were used. All other procedures were performed at 4C unless otherwise specified. The bladder fragments were homogenized in 40 ml. 50 mM. sodium potassium phosphate buffer (pH 7.4) with a Polytron PT-10–35 (Brinkman Instruments, Inc., Westbury, NY) for 60 seconds at a speed setting of 6. The initial homogenate was filtered through mesh gauze to remove connective tissue debris, and then rehomogenized with a Duall glass-glass homogenizer (Kontes Co., Vineland, NJ). The homogenate was rehomogenized with the Polytron for (60) seconds at a speed setting of 6 and again filtered through (fine) mesh gauze. An aliquot (1.5 ml.) was stored in the freezer for deoxyribonucleic acid (DNA) determination by the method of Burton[9] using calf thymus DNA as the standard.

Saturation Analysis

Saturation analyses were performed at 7 different concentrations (0.02 to 1.5 nM.) of [^{3}H]N-methylscopolamine at a constant specific activity of 84.8 Ci./mmol. (New England Nuclear Corp., Boston, MA). Total binding was determined in 1.0 ml. containing 800 μl.tissue homogenate, 100 μl, 50 mM.sodium potassium phosphate buffer (pH 7.4) and 100 ul. [^{3}H]N-methylscopolamine at varying concentrations. Nonspecific binding was determined in 1.0 ml. containing 800 μl. tissue homogenate, 100 μl. 10 μM. atropine (J. T. Baker, Phillipsburg, NJ) and 100 μl. [^{3}H] N-methylscopolamine at varying concentrations. Total binding and nonspecific binding determinations for each [^{3}H]N-methylscopolamine concentration were performed in triplicate.

The assay tubes were shaken gently on a Tectator V shaking platform (Evanston, IL) for 1 hour at room temperature, at which time a steady state was reached.[8] The binding assays were terminated by filtering on Whatman 2.4 cm. GF/B glass fiber filters (American Scientific Products, McGraw Park, IL) on a 45 well vacuum manifold (Earl Sandbek, Airville, PA). The glass filter disks were washed 4 times with 4 ml. 4C 50 mM. sodium potassium phosphate buffer (pH 7.4) under vacuum suction. The glass filter disks were placed in 10 ml. aqueous accepting scintillation fluid (Amersham-Searle Corp., Arlington Heights, IL) and allowed to equilibrate at room temperature for at least 8 hours. The scintillation vials were then counted for 10 minutes on a Searle 6890 liquid scintillation counter with an average efficiency of 40 per cent. A linear relationship between maximal specific [^{3}H]N-methylscopolamine binding and tissue concentration was demonstrated. Scatchard analysis of the data was performed using a computer assisted method.[10]

Statistical Methods

The Student t test and Wilcoxon rank sum test[11] were performed to analyze the group differences in muscarinic cholinergic receptor density.

RESULTS

Saturation experiments were performed on bladder tissue homogenates obtained from patients with ureterovesical reflux (control) and bladder exstrophy. Representative saturation curves and Scatchard plots for the binding of [^{3}H]N-methylscopolamine to bladder homogenates from control and exstrophy patients are shown in Figures 1 and 2, respectively. The binding of [^{3}H] N-methylscopolamine to control and exstrophy bladder tissue homogenates was saturable and of high affinity. The Scatchard plots were linear, suggesting a single class of [^{3}H]N-methylscopolamine binding sites. The mean equilibrium dissociation constant of [^{3}H]N-methylscopolamine binding in control (0.14± 0.027 nM.) and exstrophy (0.14 ± 0.02 nM.) bladders was similar, suggesting homogeneity of muscarinic cholinergic receptor binding sites (see table). Nonspecific binding was 10 to 15 per cent of total binding at 0.14 nM. N-methylscopolamine. The muscarinic cholinergic receptor density is expressed as fmol./ug. DNA. No significant difference was demonstrated between the mean muscarinic cholinergic receptor density in the control (1.97 ± 0.29 fmol./ug. DNA) and the exstrophy (1.44 ± 0.21 fmol./ug. DNA) bladders ($p>0.05$) (see table).

DISCUSSION

Muscarinic cholinergic receptors have been characterized in various genitourinary tissues using radioligand receptor binding methods.[12,13] This study represents the first description of muscarinic cholinergic receptors using radioligand receptor binding in human bladders. The equilibrium dissociation constants for [^{3}H]N-methyl-scopolamine binding in human, rabbit and rat bladders are similar, suggesting homogeneity of bladder muscarinic cholinergic receptor binding sites in various species.

The equilibrium dissociation constants for [^{3}H]N-methylscopolamine binding in control and exstrophied bladders were not statistically different, suggesting that the muscarinic cholinergic binding sites are not irreversibly altered by the disorder of embryogenesis resulting in exstrophy. In addition, the density of muscarinic cholinergic receptors

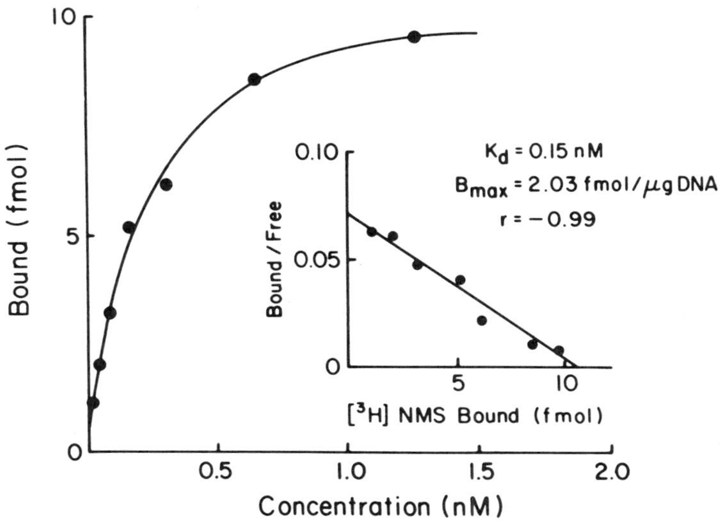

Figure 1. Representative saturation curve and Scatchard analysis of binding of [^{3}H]N-methylscopolamine to homogenates of control bladders.

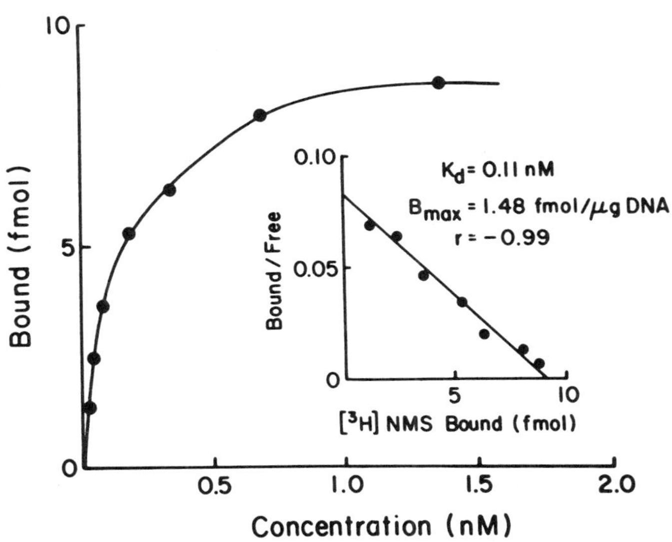

Figure 2. Representative saturation curve and Scatchard analysis of binding of [^{3}H]N-methylscopolamine to homogenates of exstrophied bladders.

in the control and exstrophied bladders was similar. Detrusor function is controlled primarily by muscarinic cholinergic receptors in association with the parasympathetic nerves originating from the sacral outflow.[7,14] The normal density of muscarinic cholinergic receptors suggests an intact autonomic innervation of the exstrophy bladder. Expression of a normal density of receptors is not unequivocal proof of normal function. Investigation of the contractile response of the bladder tissues to muscarinic agonists and electrical field stimulation studies would have provided further evidence of a normal functioning detrusor. We have subsequently compared receptor density and in vitro contactile studies in control and myelodysplastic bladders with good correlation between density and contractile response.

The anatomical, histological and biochemical properties of the exstrophied bladder are poorly understood. Further understanding of these properties may influence the surgical management of this disorder. If the exstrophy-epispadias complex results solely from failure of the cloacal membrane to be reinforced by an ingrowth of mesoderm,[16] the histological and biochemical properties of the exstrophied bladder should not be altered. However, the inherent histological properties of the exstrophied bladder remain controversial. Abnormal histological features, such as cystitis cystica and cystitis glandularis, identified in exstrophied bladders[17] may represent defective bladder development or changes secondary to chronic infection.

Our laboratory reported the first quantitative morphometric studies of the fetal and neonatal control and myelodysplastic bladders.[18] Lais et al.[19] recently reported that the neonatal exstrophy bladder is comprised of only 7% smooth muscle and following bladder neck plasty, the exstrophied bladder is comprised of 32% smooth muscle. Control bladders were comprised of approximately 50% smooth muscle. Thus, the amount of smooth muscle in the closed exstrophy bladder approaches controls. Lee et al.[20] has reported similar observations that the neonatal exstrophy blaadder is deficient in smooth muscle, but the morphometry of the closed bladder was not examined. Peppas et al.[21] has also recently shown improved cellular composition during staged reconstruction. These morphometric studies suggest that the smooth muscle deficiency is due to defunctionalization and that remodeling following closure restores the bladder to a near normal cellular composition. It is interesting to speculate that those individuals who fail to acheive continence have decreased bladder outlet resistance due to an inherently poor detrusor used in bladder neck reconstruction. Our finding of normal tissue receptors coupled with recent findings of normal cellular composition following closure provides strong evidence that most exstrophied bladders have the potential to function normally if properly managed.

REFERENCES

1. Harvard, B. M. and Thompson, G. J.: Congenital exstrophy of the urinary bladder: late results of treatment by the Coffey-Mayo method of uretero-intestinal anastomosis. J.Urol., 65: 223, 1951.
2. Marshall, V. F. and Muecke, E. C.: Functional closure of typical exstrophy of the bladder. J. Urol., 104: 205, 1970.
3. Nisonson, I. and Lattimer, J. K.: How well can the exstrophied bladder work? J. Urol., 107: 664, 1972.
4. Lepor, H. and Jeffs, R. D.: Primary bladder closure and bladder neck reconstruction in classical bladder exstrophy. J. Urol., 130: 1142, 1983.
5. Jeffs, R. D., Lakshmanan, Y., and Gearhart, J. P.: A 21-year experience with functional reconstruction in 87 consecutive patients followed from birth. Am Assoc Genito Surg Annual Meeting, 1997, Abstract.
6. Toguri, A. G., Churchill, B. M., Schillinger, J. F. and Jeffs, R. D.: Gas cystometry in cases of continent bladder exstrophy. J. Urol., 119: 536, 1978.

7. Lapides, J., Friend, C. R., Ajemian, E. P. and Sonda, L. P.: Comparison of action of oral and parenteral bethanechol chloride upon the urinary bladder. Invest. Urol., 1: 94, 1963.
8. Shapiro, E., Jeffs, R. D., Gearhart, J. P. and Lepor, H.: Muscarinic cholinergic receptors in bladder exstrophy: insights into surgical management. J. Urol., 134: 308, 1985.
9. Burton, K.: A study of the conditions and mechanism of the diphenylamine reaction for the colorimetric examination of the oxyribonucleic acid. Biochem. J., 62: 315, 1956.
10. Scatchard, G.: The attractions of proteins for small molecules and ions. Ann. N.Y. Acad. Sci., 51: 660, 1949.
11. Steel, R. G. D. and Torrie, J. H.: Principles and Procedures of Statistics: A Biochemical Approach. New York: McGraw-Hill Book Co., p. 60, 1980.
12. Levin, R. M., Staskin, D. R. and Wein, A. J.: The muscarinic cholinergic binding kinetics of the human urinary bladder. Neurourol. Urodynam., 1: 221, 1982.
13. Lepor, H. and Kuhar, M. J.: Characterization of muscarinic cholinergic receptor binding in the vas deferens, bladder prostate and penis of the rabbit. J. Urol., 132: 392, 1984.
14. Bradley, W. E. and Scott, F. B.: Physiology of the urinary bladder. In: Campbell's Urology, 4th ed. Edited by J. H. Harrison, R. F. Gittes, A. D. Perlmutter, T. A. Stamey and P. C. Walsh. Philadelphia: W. B. Saunders Co., vol. 1, chapt. 4, p. 87, 1978.
15. Gup, D. I., Baumann, M., Lepor, H. and Shapiro, E.: Muscarinic cholinergic receptors in normal pediatric and myelodysplastic bladders. J. Urol., 142: 595, 1989.
16. Muecke, E. C.: The role of the cloacal membrane in exstrophy: The first successful experimental study. J. Urol., 92: 659, 1964.
17. Culp, D. A.: The histology of the exstrophied bladder. J. Urol., 91: 538, 1964.
18. Shapiro, E., Becich, E. P. Pearlman, E. and Lepor, H.: Bladder wall abnormalities in myelodysplastic bladders: a computer assisted morphometric analysis. J. Urol., 145: 1024, 1991.
19. Lais, A., Paolocci, N., Ferro, F., Bosman, C., Boldrini, R., and Caione, P.: Morphometric analysis of smooth muscle in the exstrophy-epispadias complex. J. Urol., 156: 819, 1996.
20. Lee, B. R., Perlman, E. J., Partin, A. W., Jeffs, R. D., and Gearhart, J. P.: Evaluation of smooth muscle and collagen subtypes in normal newborns and those with bladder exstrophy. J. Urol., 156: 2034, 1996.
21. Peppas, D. S., Tchetgen, M. B., Jeffs, R. D., and Gearhart, J. P.: A quantitative histological analysis of the bladder in classical bladder exstrophy in various stages of reconstruction utilizing color morphometry. Urology, submitted for publication.

7

A QUANTITATIVE HISTOLOGICAL ANALYSIS OF THE BLADDER IN CLASSICAL BLADDER EXSTROPHY IN VARIOUS STAGES OF RECONSTRUCTION UTILIZING COLOR MORPHOMETRY

Dennis S. Peppas, Marie-Blanche Tchetgen, Benjamin R. Lee, Robert D. Jeffs, and John P. Gearhart

The Johns Hopkins Hospital
The Johns Hopkins University School of Medicine
The James Buchanan Brady Urological Institute
Department of Urology
Division of Pediatric Urology
Baltimore, Maryland 21287-2101

ABSTRACT

Purpose

The use of color image analysis to evaluate the contribution of collagen and smooth muscle in a given organ has been well described. We have recently reported on the subtypes of collagen found in normal neonatal bladders and in bladders of newborn exstrophy patients. Much debate continues in the Pediatric Urologic literature regarding the quality of the bladder in exstrophy patients. Despite what is believed to be adequate reconstruction in the exstrophy or epispadias patient, some children fail to attain a capacity required for the development of a satisfactory continence interval. The authors attempt to quantify using color morphometry, the relative amounts of collagen and smooth muscle in the exstrophic bladder at various stages of reconstruction and in newborn normal bladders

Materials and Methods

Bladder biopsies were obtained from the area just above the trigone in fifteen patients with classic bladder exstrophy and compared to sections obtained from nine neona-

The Exstrophy–Epispadias Complex, edited by Gearhart and Mathews.
Kluwer Academic / Plenum Publishers, New York, 1999.

tal bladders. All specimens were stained with Masson's Trichrome technique, highlighting the extracellular matrix as blue and smooth muscle as red. All specimens were then analyzed using color image analysis system.

Results

The collagen/smooth muscle ratio in control bladders was 0.36 ± 0.06. In those patients closed within one week of birth, the collagen/smooth muscle ratio was found to be 0.83 ± 0.46. In late closure and failed initial closure patients, the collagen smooth muscle ratio was 3.66 ± 2.3. In patients with undergoing bladder neck reconstruction the tissue/matrix ratio was 0.87 ± 0.39. These results indicate that compared with control bladder specimens, the bladder in exstrophy has an increased collagen to smooth muscle ratio. Delay in closure or failed initial closure is associated with a markedly increased COLL:SM ratio. However, once the bladder is successfully closed, the ratio returns to that level found at early closure prior to bladder neck reconstruction or augmentation.

Conclusions

Multiple factors are involved in the successful reconstruction of the exstrophy patient. Should dehiscence, chronic infections, multiple repairs (bladder closures and bladder neck reconstructions) etc. be minimized, their role determining which children will develop a satisfactory continence level may be eliminated. Therefore, the long term outcome in children born with classic bladder exstrophy may correlate most with the organization of smooth muscle and collagen as well as the type of collagen, rather than the relative amounts.

INTRODUCTION

Reconstruction of the patient with classical bladder exstrophy can lead to a high level of continence in many instances. However, continence can not be achieved without the attainment of an adequate bladder capacity. Multiple factors have been associated with the inability of the bladder to attain capacity including multiple bladder closure, recurrent infections, bladder prolapse, dehiscence, bladder calculi and vesicostomy. The exact mechanisms by which these factors impact upon eventual bladder capacity is unclear, although inflammation and fibrosis are the likely sequelae of these mechanisms.

The ability of the closed exstrophic bladder to perform the functions of filling, storage and emptying have been challenged previously.[1,2] These basic bladder functions have been questioned in the successfully reconstructed exstrophy patient, as a result of the anatomic defect seen. However, data from our institution and others suggests that the successful staged reconstruction of the exstrophic bladder can lead to continence in a high percentage of successfully closed patients.[3,4] What has yet to be completely demonstrated is the character of the bladder wall in exstrophy; anatomically and physiologically. Shapiro et al. clearly demonstrated that the bladder contains normal amounts of muscarinic receptors of bladder contractile function.[5] The quality and quantity of muscle, elastin and collagen in the bladder wall certainly plays a major role in bladder function. Nonetheless, the relative proportion of muscle and collagen in the exstrophic bladder, compared to normal control bladders is the next logical initial question to be answered in characterization of the bladder wall in this population. The purpose of this study, therefore, is to examine the relative amounts of collagen and muscle in the exstrophy bladder at various stages of reconstruction and to compare this data to the amounts of collagen and muscle found in normal bladders.

MATERIALS AND METHODS

Bladder biopsies were obtained from the floor of the bladder just above the trigone in fifteen patients with classic bladder exstrophy. Patient age at the time of biopsy ranged from 2 days to 15 years (mean 3.01 years). There were four females and eleven males in the study population. Biopsies were obtained at one of the following stages of repair: initial closure (5), reclosure (4), bladder neck reconstruction (5) and bladder neck revision (1).

Control bladder tissue was obtained from autopsy specimens from children with no genitourinary abnormalities (n = 9, age range birth through 3 days of age).

Histological slides were analyzed from the 15 exstrophy patients undergoing staged repair and from 9 control patients. Bladder biopsies obtained during surgery and autopsy were serially sectioned, mounted on glass slides and stained with Masson's trichrome technique which highlights and extracellular matrix (collagen, elastin and other connective tissue elements) as blue and smooth muscle as red.

Color images of the bladder sections were digitized using the Pulnix TMC-74 color camera mounted on an Olympus BHT-312 light microscope. Representative areas of interest within the muscularis layer and the lamina propria were digitized with the DT2871(HSI) Color Frame Grabber and processed in parallel with the DT2858 Color Frame Grabber. Digitized images were analyzed using Global Lab Color software.

RESULTS (TABLE 1)

Controls

The collagen to smooth muscle ratio in all control tissue examined was 0.36 ± 0.06 with a collagen component of 18.4 ± 5.3% of total area of muscularis layers sampled and a percentage smooth muscle of 56.2 ± 7.6%.

Early Initial Closure Patients (N = 5)

In the 5 patients that underwent bladder closure within two weeks of birth a distinct muscularis layer was identifiable, though the demarcation between longitudinal and circular muscle layers was not consistently clear. The mucosa was ulcerated with epithelial sloughing in some areas. No inflammatory cells were seen, The mean collagen to smooth muscle ratio was 0.83 ± 0.46, This ratio was significantly different from that observed in control patients. The percentage of collagen to total area sampled was 36.2 ± 9.7%, compared to 18.4 ± 5.3% in controls. The percentage of smooth muscle in these patients was 45.1 ± 11.4%, compared to 56.2 ± 7.6% in normal controls.

Late Initial Closure and Reclosure Patients (N = 4)

The mean tissue matrix ratio in these patients was 3.66 ± 2.3. This ratio was also significantly different from that observed in control patients. The percentage of collagen

Table 1.

Classification	Collagen	Muscle	C:M ratio
Controls	18.4 ± 5.3	56.2 ± 7.6	0.36 ± 0.06
Early closure	36.2 ± 9.7	45.1 ± 11.4	0.83 ± 0.46
Reclosure/late closure	62.6 ± 14.5	22.2 ± 10.8	3.66 ± 2.3
BNP, s/p BNP	36.3 ± 9.5	43.5 ± 7.4	0.87 ± 0.39

to total area sampled was 62.6 ± 14.5%, a greater than 3-fold increase over the amount quantitated in normal control bladders and a 2-fold increase over the amount quantitated in the exstrophied bladders closed within the first several months of life. The percentage of smooth muscle in these patients was 22.2 ± 10.8%, which was approximately one third of that seen in controls and half the amount estimated in the newborn exstrophies.

Bladder Neck Reconstruction and Bladder Neck Revision (N = 6)

The mean tissue matrix ratio in these patients was 0.87 ± 0.39 whereas in controls it was 0.33 ± 0.11, The relative amount of collagen in the area sampled was 36.3 ± 9.5% and the relative amount of smooth muscle was 43.5 ± 7.4%. These amounts closely parallel the amounts observed in the exstrophic bladders closed early after delivery, 36.2 ± 9.7% and 45.1 ± 11.4% respectively.

DISCUSSION

Repair of bladder exstrophy has had an increasingly favorable prognosis over the past decade Much of this success has come following the description of the staged repair as which comprises 1) early initial closure within the first days of life, with osteotomy if indicated, 2) epispadias repair to improve genital cosmesis and to enhance bladder capacity by increasing passive outlet resistance and 3) bladder neck reconstruction to achieve continence.[5] Despite the advances made the management of these patients, failures do occur. These include: failure of the initial closure, failure of the bladder to increase its capacity following epispadias repair and failure of bladder neck reconstruction.[7]

Failure of the bladder to attain adequate capacity is an important problem and an obstacle to satisfactory functional repair. It also generally precludes successful bladder neck reconstruction in which case continence cannot be achieved without augmentation or diversion.[8] Several factors have been associated with inadequate bladder capacity: an initial small bladder template, multiple bladder closures, recurrent infections, bladder prolapse, dehiscence, bladder calculi and vesicostomy. The exact mechanism by which these factors impact on eventual bladder capacity is not entirely clear, although inflammation and fibrosis are likely to be the most important. Similarly, the process by which some patients fail to develop bladder compliance despite an initial increase in capacity is not well understood. Neither is the mechanism leading to increased capacity in the time interval between epispadias repair and bladder neck reconstruction. Increased outlet resistance leading to gradual bladder volume increase at low pressure during periods of rest or at night has been offered as a possible explanation.[9] The authors postulate that the process of gradual bladder adaptation involves ultrastructural changes occurring within the exstrophic bladder from the time of birth through the various stages of repair. These changes may be viewed as maturational and their extent may determine the functional capabilities of the repaired exstrophic bladder.

Collagen, elastin and smooth muscle are components of the bladder wall which play an important role in bladder function. Collagen provides structure to smooth muscle and tensile strength to tissue. Elastin allows tissue elasticity and smooth muscle provides active contractile force. Kim et al studied the development of muscle, collagen and elastin in 15 normal human fetal bladders and noted that muscle thickness increased gradually with gestational age while the proportion of collagen to muscle decreased overall during fetal development.[10] They also found that the ratio of type I collagen, commonly present in tendon, skin and bone to type III collagen, which predominates in distensible tissue, decreased with age; and concluded that this variation was consistent with enhanced bladder compliance needed as bladder capacity was increased, In addition, they observed that the thickness and number of elastic fibers increased during gestation and postulated that elastin enhances bladder compliance during urine storage and contributes to muscle contraction in the full bladder.

Several studies have looked at the histology of exstrophic bladders both in unoperated patients and in patients who underwent anatomical closure only. Culp[11] analyzed 24 unclosed bladders. In addition to extensive urothelial changes attributed to irritative and inflammatory effects, he noted marked fibrosis of the muscle layer in 8 of the 20 specimens that had adequate muscularis for evaluation. Twelve specimens had no muscular abnormalities. His patient population varied in age from 2 weeks to 52 years, with the older patients exhibiting fibrosis more frequently.

In their series of 24 exstrophied bladders varying in age from 5 days to 52 years, Engel and Wilkinson[12] reported that 7 of the 18 adequate specimens exhibited moderate to marked fibrosis in the submucosa and in the connective tissue surrounding muscle bundles and 4 specimens showed moderate muscle bundle disorganization with variation in bundle size, shape and orientation. Rudin et al analyzed the bladders of 14 exstrophy patients. Eight were anatomically closed and had 2 or more biopsies over time and 6 were anatomically closed and subsequently underwent cystectomy for vesicoureteral reflux and infection. The ages at closure varied from 6 months to 20 years. Virtually all specimens had fibrosis and/or derangement of the muscularis propria, and these changes remained present in subsequent biopsies regardless of the age at closure.[13]

Shapiro looked at myelodysplastic bladders and demonstrated the presence of fibrotic areas infiltrated by collagen and elastin thought to cause increased wall stiffness in these noncompliant bladders.[14] Ewalt et al. used immunohistochemistry to better define the patterns of collagen and elastin deposition associated with functional alterations in the bladder[15]. They demonstrated intense localization of collagen type III and elastin within the detrusor layers of noncompliant bladders, with no significant changes of the collagen composition within the lamina propria. In their analysis of normal bladders, collagen type I and type III localized to the lamina propria and were less present in the detrusor layer, They concluded that the lamina propria is a major capacitance layer in the normal bladder, but that this capacitance layer is shifted to the detrusor layer as the muscle bundles are infiltrated by collagen. The stiffness resulting from collagen deposition would thus lead to an overall decrease in bladder compliance. Kim et al. also analyzed collagen distribution in obstructed fetal bladders.[16] In contrast, they noted that the poor contractility and decreased compliance found in these bladders correlated with an increased deposition of type I collagen around detrusor muscle bundles.

Our studies demonstrate for the first time that the ratio of collagen to smooth muscle as well as relative amounts of collagen and muscle are altered in the exstrophic bladder (Table 1). Those patients who undergo early (within the first week of life) closure have COLL:SM ratio 2.3 times that of controls (0.83/0.36). Those who undergo successful staged reconstruction have a similar collagen/smooth muscle ratio at the time of bladder neck reconstruction (0.87/0.36), The authors have previously postulated that the exstrophic bladder might behave like a defunctionalized bladder following urinary diversion[3]. However, recent data from Lais et al., suggests that unless bladder neck repair is undertaken which stimulates bladder muscle growth, the bladder will remain in a defunctionalized state.[9]

What is also unknown is the collagen/smooth muscle ratio in those children undergoing late closure or reclosure, which is twelve times that of newborn controls (3.66/0.36). It could be postulated that many of these patients will fare poorly in attaining an increase in bladder capacity and, ultimately continence with staged reconstruction. Interestingly, one half of those biopsied at the time of bladder neck reconstruction or bladder neck revision (3/6) underwent initial bladder closure at an age greater than ten days or had a failed initial closure and underwent a subsequent successful closure. One would expect that had these bladders been biopsied at the time of closure or reclosure, that they would have demonstrated a markedly increased collagen/smooth muscle ratio. However, at the time of bladder neck reconstruction or revision, these bladders demonstrated a COLL:SM ratio resembling those children undergoing early initial closure.

Few studies to date have done a quantitative analysis of collagen or smooth muscle in the exstrophic bladder at birth or at various stages of repair. Lais et al concluded in their review of nine patients at various stages of reconstruction that the ratio of smooth muscle to collagen increased, however remained below normal.[17] Lee et al., demonstrated that there was an increase in type III collagen compared to normal controls. Type III collagen has been shown to be less distensible fiber when compared to type I collagen, thereby potentially decreasing bladder compliance.[18] This type of study is crucial to our understanding of the attainment of continence following successful bladder exstrophy reconstruction. The purpose of our study was to determine the relative amounts of collagen and smooth muscle in bladder biopsies from children undergoing staged exstrophy repair utilizing quantitative histological analysis. This information was then compared to the amounts of collagen and muscle found in normal bladders. Other factors may influence ultimate success with staged reconstruction, such as the type and relative amounts of collagen, as well as the overall clinical course of the patient following exstrophy closure.

REFERENCES

1. Hollowell, J.G., Ransley, P.G.: Surgical management of incontinence in bladder exstrophy. Brit J. Urol., 68:543, 1991.
2. Hollowell, J.G., Hill, P.D., Duffy, P.G., Ransley, P.G.: Bladder function and dysfunction in exstrophy and epispadias. Lancet 338:926, 1991.
3. Peppas, D.S., Gearhart, J.P., Jeffs, R.D.: Genitourinary reconstruction in classic bladder exstrophy: Results of a seventeen year experience in a single institution. J Urol (submitted)
4. Husmann, D.A., McLorie, G.A., Churchill, B.M.: Closure of the exstrophic bladder: An evaluation of factors leading to its success and its importance on urinary continence. J Urol., 142:522, 1989.
5. Shapiro, E., Jeffs, R.D., Gearhart, J.P., Lepor, H.: Muscarinic cholinergic receptors in bladder exstrophy: Insights into surgical management. J Urol 134:308, 1985.
6. Jeffs, R.D., Charrois, R., Many, M., Juriansz, A.R.: Primary closure of the exstrophied bladder. Current controversies in Urologic Management. Edited by R. Scott. Philadelphia: W.B. Saunders Co., pp. 226–242, 1972.
7. Gearhart, J.P.: Failed bladder exstrophy repair - Evaluation and Management. Urol Clin North Amer. 18:4, 1991.
8. Gearhart, J.P.: Bladder neck reconstruction in the incontinent child. Dialogues Pediatr Urol. 10 October 1997.

9. Lais, A., Padiocci, N., Ferro, F., Bosman, C., Boldrini, R., Caione, P.: The fate of smooth muscle in exstrophy-epispadias. Abstract American Academy of Pediatrics Urology Section, San Francisco, CA Oct 15 1995.
10. Kim K., Kogan, B., Massad, C., Huang, Y-C.: Collagen and Elastin in the normal fetal bladder. J Urol., 146:524, 1991.
11. Culp, D.: The histology of the exstrophied bladder. J Urol. 91:5, 1964.
12. Engel, R., Wilkinson, H.: Bladder Exstrophy. J Urol. 104:699, 1970.
13. Rudin, L., Tannenbaum, M., Lattimer, J.: Histologic analysis of the exstrophied bladder after anatomical closure. J Urol. 108:802, 1972,
14. Shapiro, E., Becich, M. Perlman, E., Lepor, H.: Bladder wall abnormalities in myelodysplastic bladders: a computer assisted morphometric analysis. J Urol. 145:1024, 1991.
15. Ewalt, D., Howard, P., Blyth, B., Snyder, III, H., Duckett, J., Levin, R., Macarak, E.: Is lamina propria matrix responsible for normal bladder compliance? J Urol. 148:544, 1992.
16. Kim, K., Kogan, B., Massad, C., Huang, Y-C.: Collagen and elastin in the obstructed fetal bladder. J Urol. 146: 528, 1991.
17. Lais, A., Paolocci, N., Ferro, F., Bosman, C., Boldrini, R., Caione, P.: Morphometric analysis of smooth muscle in the exstrophy-epispadias complex. J Urol., 156:819, 1996.
18. Lee, B.R., Perlman, E.J., Partin, A.W., Jeffs, R.D., Gearhart, J.P.: Evaluation of smooth muscle and collagen subtypes in the normal and bladder exstrophy patient. J Urol., 156: 2034, 1996.

8

ANALYSIS OF SMOOTH MUSCLE AND COLLAGEN SUBTYPES IN THE NORMAL NEWBORN AND BLADDER EXSTROPHY PATIENT

Benjamin R. Lee, Elizabeth J. Perlman, Alan W. Partin, Robert D. Jeffs, and John P. Gearhart*

Departments of Urology and Pathology
The Johns Hopkins School of Medicine
Baltimore, Maryland

ABSTRACT

Purpose

Many patients who undergo newborn bladder exstrophy closure, subsequent epispadias repair and later bladder neck reconstruction become completely continent, yet complications can occur. After a successful initial exstrophy closure and later epispadias repair, some patients may fail to gain sufficient capacity for bladder neck reconstruction. Others may fail to gain satisfactory capacity and continence after bladder neck reconstruction. In an attempt to understand the pathogenesis of these failures, the authors first compared bladder biopsies from normal neonatal bladders to newborn exstrophy bladders.

Materials and Methods

Bladder biopsies were taken from midline of the bladder wall just above the base of the trigone in twelve newborn exstrophy patients and compared to bladder sections from nine expired neonates. All bladder specimens were stained with monoclonal antibodies against Types I, III or IV collagen. A subset were further stained with Masson's Trichrome

* Contact author: Division of Pediatric Urology—Marburg 149, The James Buchanan Brady Urological Institute, The Johns Hopkins Hospital, Baltimore, MD 21287-2101

The Exstrophy–Epispadias Complex, edited by Gearhart and Mathews.
Kluwer Academic / Plenum Publishers, New York, 1999.

to define the extracellular matrix. All specimens were then analyzed using a color digital image analysis system.

Results

On initial examination of the extracellular matrix, there was an increase in collagen:smooth muscle ratio from 0.38 in controls to 1.2 in newborn exstrophy patients. This ratio was composed of both an increase in collagen as well as a decrease in smooth muscle. The collagen component of the extracellular matrix was then further defined to quantitate how much of each collagen type (I, III & IV) was being deposited.

The authors then examined the ratio of collagen type to total collagen sampled. Compared to control bladders, there was not a statistical difference in the amount of types I or IV in the newborn exstrophy bladder at time of initial closure. Interestingly however, there was a three fold increase in type III collagen, from 0.14+/-.05 to 0.46 +/-0.2 ($p \ll 0.001$) in the neonatal control bladders vs. newborn exstrophy bladders, respectively.

Conclusions

This alteration in collagen makeup may represent an earlier developmental stage of the exstrophied bladder at birth which then remodels and changes after successful initial closure. Further studies are underway to examine the collagen composition of bladders at the time of bladder neck reconstruction and failed closures as well as those undergoing bladder augmentation.

INTRODUCTION

The achievement of adequate bladder capacity in the exstrophy patient plays a key role towards eventual urinary continence. Multiple factors have been associated with the inability of the exstrophy bladder to attain capacity including multiple bladder closures, recurrent infections, bladder prolapse, dehiscence, bladder calculi and vesicotomy.[1] The exact mechanisms through which these factors influence bladder development is unclear, although inflammation and fibrosis are the likely sequelae of these mechanisms.

Great strides towards achieving a successful repair have included early closure with better preservation of bladder epithelium and underlying muscle, technical improvement in initial closure and osteotomy when necessary, and management of reflux after initial closure and at the time of bladder neck reconstruction. Also, development of staged repair allows for bladder growth and increased volume before bladder neck plasty for incontinence, including initial epispadias repair prior to bladder neck plasty to encourage bladder adaptation at low pressure to increased volumes.[2]

Collagen, smooth muscle and elastin compose important cellular and histological elements responsible for bladder function. Types I and III collagen provide the tensile strength within the bladder.[8] Type I collagen is also found in skin, bone, ligaments and tendon; it accounts for 90% of body collagen and is the major collagen fiber in the intermuscular bundle space. Type III collagen is present in the smooth muscle of the detrusor, uterus, small intestine and other internal organs; it is found within the muscular bundle between individual muscle fibers.

An analysis of these collagen subtypes may elucidate why some newborn exstrophic bladders grow to achieve a significant capacity while others fail to attain an adequate bladder capacity and must eventually undergo an augmentation procedure.

MATERIALS AND METHODS

Patients

Bladder biopsies were obtained from twelve newborn infants with bladder exstrophy undergoing initial closure. Patient age at the time of biopsy ranged from one day to twenty days (mean = seven days). There were three males and nine females in the study population. Biopsies were all obtained from the midline of the bladder wall just above the base of the trigone at initial closure during the newborn period.

Control bladder tissue was obtained from the bladder floor in nine expired neonates at autopsy (mean age = five days). Cases were screened to exclude those who had been catheterized during their hospital course or had any genitourinary malformation.

Histological slides were analyzed from the twelve exstrophy patients undergoing staged repair in the newborn period and from the nine control patients. Bladder biopsies obtained during surgery and at autopsy were paraffin-embedded and serially sectioned, mounted on poly-L-lysine coated slides and stained with Type I, III and IV anticollagen antibodies (BioGenex Laboratories) and with Masson's trichrome stain which stains the extracellular matrix collagen as blue and smooth muscle as red.

Color Image Analysis

Color images of the bladder sections were digitized using a Pulnix TMC-74 color camera mounted on an Olympus BHT-312 light microscope. Representative areas of interest within the muscularis layer and the lamina propria were digitized with the DT2871(HSI) Color Frame Grabber and processed in parallel with the DT2858 Color Frame Grabber Image Processor. A histogram of the area of interest was generated with a 256 pseudo color abscissa vs. the pixel (picture elements) count ordinate. Thresholds were set on the color axis correlating to collagen, smooth muscle and background. The areas under the histogram curve were then totaled to quantitate each component. Images were analyzed using software written in Microsoft C. All statistical comparisons were made using the unpaired t-test.

RESULTS

Controls

The extracellular matrix in the nine neonatal control bladders exhibited a predominance of smooth muscle, 51% +/-11% with only 19% +/-1% collagen. The collagen to smooth muscle ratio was therefore calculated to be 0.38.

Examination of the collagen subtypes revealed that the ratio of type I collagen to total collagen sampled was 0.73 +/- 0.3. Lesser amounts of type III and IV collagen were detected, 0.14 +/- 0.05, and 0.13 +/- 0.15, respectively. As shown in Figure 1A, there is very little histochemical staining of type III collagen among the muscle bundles of the muscularis mucosa layer. Similar paucity of type III collagen was located within the lamina propria.

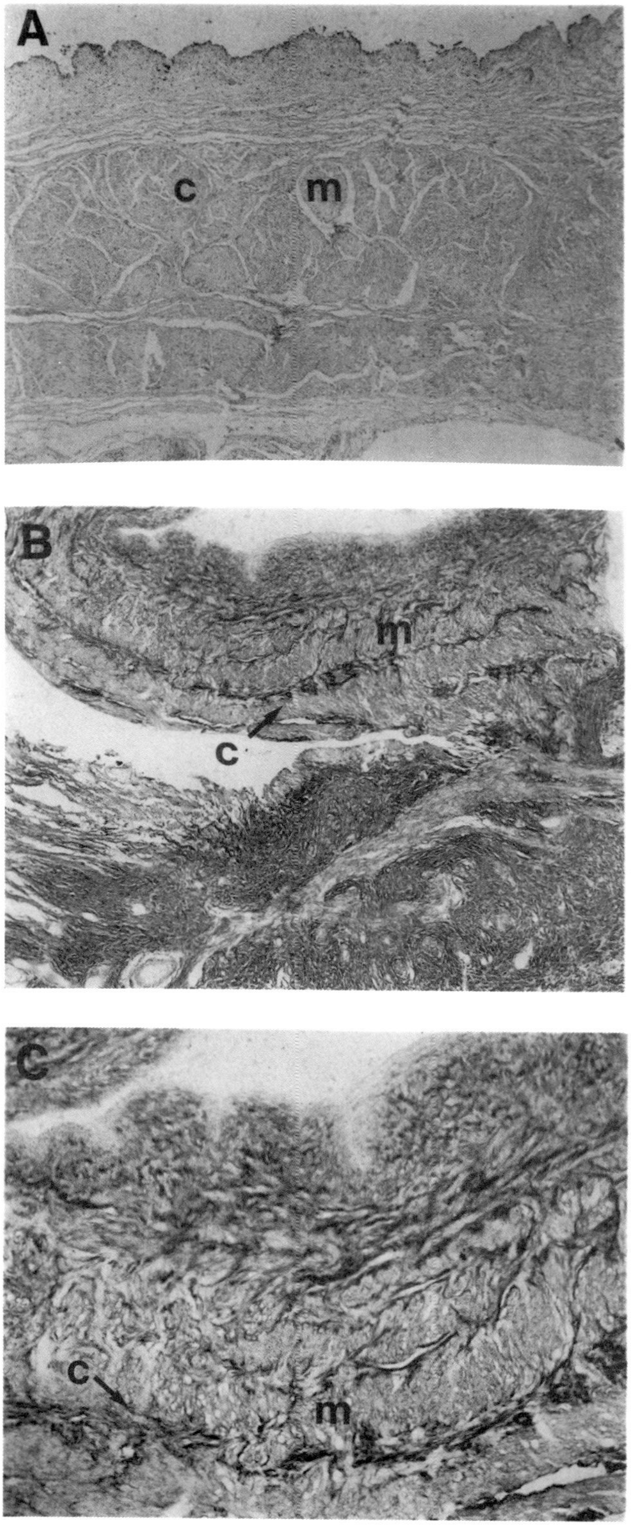

Figure 1. Type III collagen. Photomicrographs of type III collagen antibody-stained tissue—Control neonatal bladder, 40x (A) and newborn exstrophy bladders, 40x (B) and 100x (C). The dark pigmented histochemical stain is type III collagen deposition (c); smooth muscle has also been labeled (m) accordingly. Figure 1A shows little type III collagen staining in neonatal control bladders. Figure 1B demonstrates the presence of type III collagen between the muscle fascicles and throughout the muscularis mucosa which is also shown in higher magnification in Figure 1C.

Table 1. Collagen:smooth muscle ratio of neonatal control bladders vs. newborn exstrophy bladders. A three fold increase in tissue matrix ratio is made up of both an increase in collagen and a decrease in smooth muscle

	Neonatal control	Newborn exstrophy
Collagen	19 +/- 1%	39 +/- 13%
Smooth muscle	51 +/- 11%	32 +/- 11%
Tissue matrix ratio (Coll:SM)	0.38	1.2

Newborn Exstrophy Bladders

In this population, there were differences in the makeup of the extracellular matrix. The amount of smooth muscle was found to be only 32% +/- 11% compared to controls (51% +/-11%) whereas the amount of collagen increased relative to controls to 39% +/- 13% from 19%+/1%. The differences in tissue matrix ratio (collagen : smooth muscle) to 1.2 in newborn exstrophy bladders is therefore composed of both differences in collagen and in smooth muscle (Table 1).

In newborn exstrophy bladders stained with collagen antibodies, there was a significant amount of type I and III collagen deposited within the muscularis mucosa layer. Significant amounts of type III collagen was detected between the muscle fascicles, as demonstrated by the dark pigmented histochemical stain in Figures IB and 1C. Interestingly, the ratio of type III collagen to total collagen sampled in the newborn exstrophy bladder increased to 0.46 +/- 0.2. This ratio was a three fold increase compared to neonatal controls ($p<<0.001$) (Figure 2). Finally, the amount of type IV collagen was found to be only 0.09 +/- 0.08. There was no statistically significant difference in type IV collagen

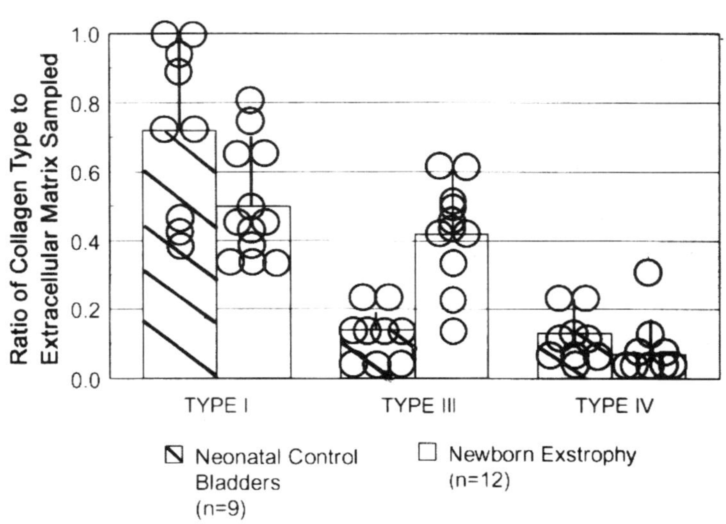

Figure 2. Graph of neonatal control bladders (crosshatched pattern) compared to newborn exstrophy bladders (clear pattern). Circles represent the data points for the nine control bladders and twelve exstrophy bladders. There is no statistical difference comparing type I collagen between the bladder types. However, there was a three fold increase in amount of type III collagen in newborn exstrophy bladders compared to controls.

content in neonatal control bladders compared to the exstrophy group. In fact the meager amount of collagen detected was not above background values of variance of the histochemical stain. Therefore, we do not believe type IV collagen exists within the bladder in any significant amount.

DISCUSSION

Achievement of adequate bladder capacity and continence in the child with bladder exstrophy poses a most formidable challenge. Staged functional repair involves primary bladder closure, correction of incontinence and reflux, and epispadias repair or female genital reconstruction. If bladder capacity proves inadequate or incontinence persists, bladder neck reconstruction and augmentation, or continent urinary diversion may be considered. Why some patients gain adequate bladder capacity and can progress to bladder neck repair and continence while other patients never develop an adequate bladder capacity and must undergo augmentation and an outlet procedure remains unclear.

The authors demonstrate for the first time a change in the structure of the newborn exstrophy bladder, that is, an increase in Type III collagen compared to neonatal controls. Type III collagen may play a key role in understanding the differences in bladder capacity.

It has been postulated that a relationship exists between the percentage of smooth muscle present in the detrusor and the quality of vesical contraction. A relative increase in collagen is likely to reduce that quality of the contraction.[3] Baskin et. al. examined the effect of physical forces on bladder smooth muscle and urothelium and found that physical strain of bladder wall cells in vitro can alter secretion of connective tissue elements.[4] There was a significant increase in collagen type III and fibronectin synthesis in stressed cells when compared to unstressed cells. Type III collagen may be a less distendable fiber, which decreases the compliance of the bladder. It is somewhat surprising that exstrophy bladders would have an increase in collagen content in the absence of outlet obstruction. However, during fetal life as maturation and growth of the intraabdominal viscera occurs the bladder template may be intermittently extruded and stretched which could be a stimulus for either the deposition or persistence of Type III collagen.

The role that type III collagen plays in the story of bladder compliance is still being defined. Ewalt et al. used two-dimensional gel electrophoresis to show that type III collagen increases disproportionately to type I collagen in the non-compliant neurogenic bladder. They suggested that type III collagen accumulation within muscle bundles may contribute significantly to trabeculation and the mechanical properties of the non-compliant bladder.

Landau described ultrastructural changes that occur in the wall of dysfunctional bladders and concluded that the poor storage function of poorly compliant bladders is secondary to an alteration in the connective tissue content of the bladder wall.[5] Their results show abnormal bladders to have 35.19 +/- 2.84% connective tissue, which is similar to our findings in the newborn exstrophy bladder of 39 +/- 13% collagen. The difference in collagen:smooth muscle ratio of 0.6 vs. that found in our study in exstrophy bladders of 1.2 may reflect a different stage of collagen remodeling taking place. The exstrophy bladder prior to birth is not yet exposed to the distorting pressure and volume effects of filling and emptying urine. Kim, Kogan et. al. have shown that Type III collagen is present *in utero*, and decreases to baseline levels just before birth whereas Type I does just the opposite.[6] Infants with bladder exstrophy may represent an earlier stage of collagen development in a bladder which has not been subjected to the same pressure and volume forces as normal bladders.

Peppas et al. examined the tissue matrix ratio of the exstrophic bladder at different stages of repair.[7] These results corroborate our findings of an increase in the collagen:smooth muscle ratio of 2.5 times that of controls in patients who undergo early closure. Interestingly, in children undergoing late closure or reclosure, the tissue matrix ratio increased to twelve times that of controls. Peppas postulated that these patients with an increased tissue matrix ratio will fare poorly in attaining a sufficient increase in bladder capacity and ultimately continence with staged reconstruction. Furthermore, it was shown that patients who undergo successful staged reconstruction have a return to a tissue matrix ratio at time of bladder neck reconstruction similar to that of the newborn period.

Further studies are currently underway to delineate collagen subtypes in those bladders who achieve adequate capacity for bladder neck reconstruction, bladders who fail initial or secondary closure and those exstrophy bladders undergoing augmentation cystoplasty. Hopefully, these studies may help us to predict which exstrophy bladders may proceed with staged reconstruction and which may need adjunctive surgical procedures for reconstruction of the bladder.

REFERENCES

1. Gearhart JP: Failed bladder exstrophy repair - evaluation and management. Urol. Clin. North Amer. **18:**4, 1991.
2. Gearhart JP, Peppas DS and Jeffs RD. The failed exstrophy closure: strategy for management. Brit J Urol. **71:**217–22, 1993.
3. Susset JG, Servot-viguier D, Lamy F, Madernas P, and Black R. Collagen in 155 human bladders. Invest. Urol. **16(3):**204–206, 1978.
4. Baskin L, Howard PS and Macarak E. Effect of physical forces on bladder smooth muscle and urothelium. J Urol. **150:**601–607, 1993.
5. Landau EH, Jayanthi VR, Churchill BM, Shapiro E, Gilmour RF, Khoury AE, Macarak EJ, McLorie GA, Steckler RE and Kogan BA. Loss of elasticity in dysfunctional bladders: Urodynamic and histochemical correlation. J Urol., **152:**702–705, 1994.
6. Kim KM, Kogan BA, Massad CA and Huang YC. Collagen and elastin in the normal fetal bladder. J Urol. **146:**524–527, 1991.
7. Peppas DS, Tchetgen MB, Jeffs RD and Gearhart JP. A quantitative histological analysis of the bladder in classical bladder exstrophy in various stages of reconstruction utilizing color morphometry. Submitted for publication, 1996.
8. Macarak EJ, Howard PS and Lally ET. Production and characterization of a monoclonal antibody to human type III collagen. J Histo Cyto. **34(8):**1003–1011, 1986.

9

A LARGE ANIMAL MODEL OF BLADDER EXSTROPHY

Observations on Bladder Smooth Muscle and Collagen Content and Function

Bruce L. Slaughenhoupt[1*] and John P. Gearhart[2]

[1]Division of Urology, Department of Surgery
University of Louisville
Louisville, Kentucky 40292
[2]Brady Urological Institute
Johns Hopkins Hospital
Baltimore, Maryland, 21287

ABSTRACT

Purpose

While most patients who undergo bladder exstrophy closure develop sufficient bladder capacity, others do not. To discover the factors which lead to these differences, we compare bladder biopsies from normal neonatal sheep to those of sheep which have undergone creation of bladder exstrophy. We then compared these results with those of a previous study in which we compared bladder biopsies from human neonatal bladders to newborn exstrophy bladders.

Material and Methods

In earlier work from our laboratory, bladder specimens from 9 normal neonatal human bladders were compared to 12 newborn exstrophy bladders. We demonstrated an increase in collagen and a decrease in smooth muscle in the exstrophy bladders as compared

* Contact author: Bruce L. Slaughenhoupt, Division of Urology, Department of Surgery, University of Louisville, Louisville, Kentucky 40292; Telephone No. (502) 852-5447; Fax No. (502) 852-8915

The Exstrophy–Epispadias Complex, edited by Gearhart and Mathews.
Kluwer Academic / Plenum Publishers, New York, 1999.

to the normal bladders. We also demonstrated a threefold increase in Type III collagen in the exstrophy bladders when compared to the controls. Now we proceed with a similar analysis between normal and exstrophic lamb bladders developed via a previously published animal model of bladder exstrophy. Bladder specimens from 10 newborn control lambs and 7 lambs with bladder exstrophy were stained with Maison trichrome as well as monoclonal antibodies to Type I and Type III collagen. Specimens were then analyzed via a color digital image analysis system.

Results

In the sheep model, there was a significant increase in collagen versus smooth muscle in the exstrophy bladders when compared to the controls ($p<0.05$). This finding is similar to that seen in human exstrophy bladders. However, unlike in the human tissue, we did not find a significant difference in the ratio of Type I and Type III collagen between the two groups.

Conclusion

The exstrophy bladder specimens from the fetal lamb model reveal a similar increase in collagen versus smooth muscle, as do the specimens from our exstrophy patients. They do not, however, demonstrate the same increase in collagen Type III, as do the patient specimens.

INTRODUCTION

The achievement of an adequate bladder capacity in the exstrophy patient plays a large role toward eventual urinary continence. Multiple factors have been associated with the inability of the patient with bladder exstrophy to attain continence, including small bladders, multiple bladder closures, recurrent infections, and bladder calculi.[1] The exact mechanisms through which these factors affect bladder development are unclear. To better understand how various factors actually enhance or inhibit bladder development, as well as to assess better techniques for bladder closure and pubic bone approximation, we developed a large animal model of bladder exstrophy.[2] Specimens from exstrophied and control sheep bladders are now analyzed for smooth muscle and collagen content and subtypes.

MATERIALS AND METHODS

Ewes with fetuses at 70 to 80 days of gestation (term - 150 days) were used in this study. The surgical technique used to create the exstrophy bladders as well as the control bladders is detailed in a previous publication.[2] Exstrophied bladders were created primarily in female fetuses. However, due to the large number of male fetuses, we also performed exstrophy creating surgery on several male fetuses. In the male fetuses that underwent surgical manipulation, a paramedian abdominal incision was made. The remainder of the procedure was carried out, as with the female fetuses.

All fetuses were then allowed to go to term and undergo spontaneous vaginal delivery. All lambs that survived surgery, gestation, and delivery were then sacrificed, weighed and autopsied. Bladders were harvested and weighed before sectioning for formalin preservation or rapid freezing in liquid nitrogen. Specimens of the mature bladders were then stained with Masson trichrome, as well as anticollagen monoclonal antibodies against sheep Type I and Type III collagen. Color images of the bladder sections were digitized using a Pulnix TMC-74 color camera mounted on an Olympus BHT-312 light microscope. Representative areas of interest within the muscularis layer and the lamina propria were digitized with the DT-2871 Color Frame Grabber and processed in parallel with the DT-2858 Color Frame Grabber image processor. Digitized images were then analyzed using software from the Global Lab System.

RESULTS

Fifteen female sheep underwent creation of bladder exstrophy. Of these 15, 6 were born live and remained viable until sacrificed. Three were born live but died soon after delivery and 6 experienced in utero fetal demise. Eight male fetuses underwent creation of bladder exstrophy. Three of these were born live and survived until they were sacrificed. Five experienced in utero fetal demise. Sixteen male fetuses underwent sham surgery. Ten of those were born live and remained alive until they were sacrificed. Three experienced fetal demise. Three were born live but then died shortly after delivery (Table 1).

Bladder specimens from 10 newborn control lambs and 7 lambs with bladder exstrophy were stained with Maison trichrome, as well as monoclonal antibodies against Type I and Type III collagen.

Compared to control bladders, the exstrophy bladders had a significant increase in collagen versus smooth muscle. The exstrophied bladders contained 32.2 percent smooth muscle and 43.4 percent collagen. The control bladders had 42.6 percent smooth muscle and 32.1 percent collagen. When the collagen subtypes were further delineated by staining with monoclonal antibodies to Type I and Type III collagen, there was no significant increase in the amount of Type III collagen in the exstrophied bladders. Rather they contained 38.6 percent Type I collagen and 33.8 percent Type III collagen, while the control bladders contained 45.8 percent Type I collagen and 35.1 percent Type III collagen (Table 2).

DISCUSSION

Classic bladder exstrophy is a congenital birth defect which is said to range in incidence between 1 in 25,000 live births and 1 in 33,000 live births.[3,4] It is known to occur more commonly in males than females by ratios that range between 2.3 to 1 and 6 to 1.[5,6] It is also known that a person with the exstrophy/epispadias complex has a 1 in 70 chance of having a child with bladder exstrophy.[7]

We have reported elsewhere on our evaluation of smooth muscle and collagen subtypes in the normal newborn child, as well as in the child born with bladder exstrophy.[8] In that report, we demonstrated an increase in collagen and a decrease in smooth muscle in

Table 1. Results

	Female exstrophy	Male control	Male exstrophy
n	15	16	8
Live births	6 (40%)	10 (62%)	3 (38%)
Fetal demise	6 (40%)	3 (19%)	5 (62%)
Post partum death	3 (20%)	3 (19%)	0

Table 2. Bladder composition

Sm. muscle %	Collagen %	I	III
Control bladders			
38.8	49.0	17.5	42.7
52.9	28.2	25.8	46.8
62.8	18.5	35.4	23.1
39.8	23.9	33.4	24.0
35.1	40.7	38.2	36.3
49.7	34.0	41.9	46.2
35.5	37.1	27.1	39.4
20.8	28.6	36.4	32.1
37.3	35.6	36.4	37.7
53.5	25.2	42.9	22.9
42.6	32.0	45.8	35.1
Exstrophied bladders			
45.0	24.0	39.2	20.9
33.7	48.9	49.5	40.8
23.5	56.0	42.6	35.8
26.1	38.2	18.4	42.5
30.0	47.6	32.2	39.0
26.9	47.1	34.0	27.8
40.4	42.1	54.2	30.0
32.2	43.4	38.6	33.8

the exstrophy bladders as compared to the normal bladders. Further, a three-fold increase in Type III collagen was demonstrated in the exstrophy bladders when compared to the controls. It was postulated that this variation in collagen composition in the exstrophy bladder may represent an earlier developmental stage of the bladder at birth due to the lack of normal filling and emptying which the normal bladder experiences.

We now analyze our animal model of bladder exstrophy to see if similar results are found. Again, a significant increase in the percent of collagen in the exstrophied bladders as opposed to smooth muscle was noted. However, the animal model did not demonstrate the same increase in Type III collagen as seen in the clinical study. This variation between the animal model and the clinical study most likely reflects that in the animal model, the bladders undergo exstrophy creation at mid-gestation. Hence, the bladder has been undergoing sequential filling and emptying until mid-gestation. This differs from the human samples in which the exstrophic condition is present from the beginning of gestation.

CONCLUSIONS

The exstrophy bladder from the fetal lamb model demonstrates an increase in collagen as opposed to smooth muscle. This is similar to findings reported in patients with classic bladder exstrophy. The exstrophied bladder specimens from the lamb model do not, however, demonstrate an increase in Type III collagen as is seen in clinical specimens. This may be due to the fact that the animal model allows the bladder to undergo normal development until mid-gestation.

REFERENCES

1. Gearhart, J.P. Failed bladder exstrophy repair. Evaluation and Management. Urol. Clin. North Am., **18**:687, 1991.

2. Slaughenhoupt, B.L., Chen, C.J., Gearhart, J.P.: Creation of a model of bladder exstrophy in the fetal lamb. J. Urol., **156**:816, 1996.
3. Yang, P., Khoury, M.J., Stewart, W.F., Beaty, T.H., Chee, E., Beatty, J.C., Diamond, E.L., Gordis, L.: Comparative epidemiology of selected midline congenital abnormalities. Genet. Epidemiol. **11**(2):141, 1994
4. Kockum, C.C., Hansson, E., Stenberg, A., Svensson, J., Malmfors, G. Bladder exstrophy in Sweden—a long-term follow-up study. Eur. J. Pediatr. Surg. **6**(4):208, 1966.
5. Bennett, A.H.: Exstrophy of the bladder treated by ureterosigmoidostomies. Long term evaluation. Urology **2**:165, 1973.
6. Ives, E., Coffey, R., Carter, C.O.: A family study of bladder exstrophy. J. Med. Genet., **17**:139, 1980.
7. Shapiro, E., Lepor, H., Jeffs, R.D.: The inheritance of the exstrophy-epispadias complex. J. Urol., **132**:308, 1984.
8. Lee, B.R., Perlman, E.J., Partin, A.W., Jeffs, R.D., Gearhart, J.P.: Evaluation of smooth muscle and collagen subtypes in normal newborns and those with bladder exstrophy. J. Urol.,**156**(6):2034, 1996.

10

TISSUE ENGINEERING TECHNIQUES FOR CLOSURE OF BLADDER EXSTROPHY

An Experimental Animal Model

Anthony Atala

Children's Hospital and Harvard Medical School
Boston, Massachusetts

The prenatal diagnosis of patients with the exstrophy-epispadias complex is now more prevalent. Prenatal ultrasonography allows for a thorough survey of fetal anatomy. The absence of bladder filling, a mass of echogenic tissue of the lower abdominal wall or a low set umbilicus during prenatal sonographic examination may suggest the diagnosis of bladder exstrophy. These findings and the presence of intraluminal intestinal calcifications suggest the presence of a cloacal malformation.

The natural consequence of the evolution in prenatal diagnosis led to the use of intervention before birth to reverse potentially life threatening processes. However, the concept of prenatal intervention itself is not limited to this narrow group of indications. A prenatal, rather than a postnatal diagnosis of exstrophy, may be beneficial under certain circumstances. There is now a renewed interest in performing a single-stage reconstruction in some patients with bladder exstrophy. Limiting factors for following a single or multiple-stage approach may include the findings of a small, fibrotic bladder patch without either elasticity or contractility, or a hypoplastic bladder. Even if the bladder in a particular patient is deemed adequate in size, other problems may prevent closure, such as a deficiency of skin or an abnormally large pubic diastasis.

There are several strategies which may be pursued, using today's technological and scientific advances, which may facilitate the future management of patients with bladder exstrophy. Having a ready supply of urologic associated tissue for surgical reconstruction at birth may be advantageous. Theoretically, once the diagnosis of bladder exstrophy is confirmed prenatally, a small bladder and skin biopsy could be obtained via ultrasound guidance. These biopsy materials could then be processed and the different cell types expanded in vitro. Using tissue engineering techniques developed at our center and described previously, reconstituted bladder and skin structures in vitro could then be readily available at the time of birth for a one-stage reconstruction, allowing for an adequate anatomic and functional closure.

The Exstrophy–Epispadias Complex, edited by Gearhart and Mathews.
Kluwer Academic / Plenum Publishers, New York, 1999.

Toward this end, we conducted a series of experiments using fetal lambs. Bladder exstrophy was created surgically in ten 90–95 day gestation fetal lambs. The lambs were randomly divided into two groups of five. In group I, a small fetal bladder specimen was harvested via fetoscopy. The bladder specimen was separated and muscle and urothelial cells were harvested and expanded separately under sterile conditions in a humidified 5% CO_2 chamber, as previously described. Seven to ten days prior to delivery, the expanded bladder muscle cells were seeded on one side and the urothelial cells on the opposite side of a 20 cm^2 biodegradable polyglycolic acid polymer scaffold. After delivery, all lambs in group I had surgical closure of their bladder using the tissue engineered bladder tissue. No fetal bladder harvest was performed in the group II lambs, and bladder exstrophy closure was performed using only the native bladder. Cystograms were performed 3 and 8 weeks after surgery. The engineered bladders were more compliant (p=.01) and had a higher capacity (p=.02) than the native bladder closure group. Histologic analysis of the engineered tissue showed a normal histologic pattern, indistinguishable from native bladder at 2 months.

In another experiment, ten 90–95 day gestation fetal lambs underwent a skin biopsy over the lumbar area via fetoscopy. The specimens were dissected and dermal fibroblasts and keratinocytes were harvested and expanded separately. Ten days before delivery, the fibroblasts and keratinocytes were seeded on one side of the biodegradable polymer scaffold. After delivery, the autologous engineered skin was implanted over a large paramedian excisional skin defect in all 10 lambs. A second skin defect region received an absorbable polymer without cells as a control. Excisional biopsies of both implant sites and normal skin were obtained weekly for eight weeks. Fetal and neonatal survival rates were 100%. The engineered graft sites showed a higher level of epidermal organization and decreased dermal scarring than the acellular grafts.

Both of the studies described above show that the potential for replicating this technology in humans is possible. Certainly, other tissues, such as cartilage, corpora cavernosa and bowel can also be harvested and expanded in the same manner. Similar studies addressing these tissues are now in progress in our laboratory.

In addition to being able to manage the bladder exstrophy complex in utero with tissue engineering techniques, one could also manage patients after birth in a similar manner, whenever a prenatal diagnosis is not assured. In these instances, tissue biopsies could be obtained at the time of the initial surgery. Different tissues could be harvested and stored for future reconstruction, if necessary. A tissue bank for exstrophy complex patients could preserve different cell types indefinitely.

In addition to having an exstrophy complex tissue bank serve as a repository of cells for future tissue reconstitution, it could also serve as a resource to elucidate the cellular, molecular and genetic mechanisms required for the development and future prevention of these anomalies.

11

SURGICAL PRINCIPLES IN THE PRIMARY CLOSURE OF BLADDER EXSTROPHY

John P. Gearhart and Ranjiv Mathews

Division of Pediatric Urology
James Buchanan Brady Urological Institute
Johns Hopkins Hospital and School of Medicine
Baltimore, Maryland

INTRODUCTION

The functional closure of bladder exstrophy requires initial secure bladder closure followed by epispadias repair and finally bladder neck reconstruction for continence.[1] Good functional and cosmetic results are best obtained with closure and follow-up performed by a team dedicated to the management of this very complex anomaly.

INITIAL EVALUATION

Prenatal ultrasonographic diagnosis[2] allows for appropriate parental counseling, diagnosis of associated conditions and referral to a tertiary care center for delivery and immediate post-natal management. Delivery room care, while directed to stabilization of the newborn infant, should include appropriate measures to protect the delicate exposed bladder mucosa.

The timing of surgical intervention is dependent on the overall condition of the infant, the presence of associated anomalies that may be corrected at the time of bladder closure or may preclude bladder closure, the size of the bladder template, the extent of pubic diastasis and the need for osteotomy. If primary closure is delayed (>6 months), combined exstrophy closure and epispadias repair following testosterone stimulation may be considered.

The Exstrophy–Epispadias Complex, edited by Gearhart and Mathews.
Kluwer Academic / Plenum Publishers, New York, 1999.

PRIMARY BLADDER CLOSURE

A planned and carefully executed surgical procedure are keys to a successful outcome. The initial incision includes the umbilicus and bladder plate and extends distal to the verumontanum in the male or the vaginal orifice in the female. Incision of the urethral plate is performed only when urethral groove length is inadequate.[1] Additional penile length may be gained by dissecting the corpora from their attachments to the suspensory ligaments and the inferior pubic ramii.[1] When mucosal plate transection is required, thin rotational flaps may be employed to cover the bare corporae.

Excision of the umbilicus is performed and the detrusor is separated from the rectus sheaths bilaterally. Extraperitoneal dissection is possible by initially separating the peritoneum above the umbilicus. Detachment of the urogenital diaphragm from the inferior ramus of the pubis is essential to allow the bladder and posterior urethra to drop deeply into the pelvis. At the completion of bladder closure, the epispadiac urethral meatus located at the base of the penis should accommodate a 14 French sound without difficulty. The posterior urethral closure should provide a degree of outlet resistance to promote bladder growth and prevent prolapse, without causing urinary retention and upper tract deterioration. A second layer of surrounding tissue may be used to further reinforce the bladder neck and posterior urethra. Paraexstrophy flaps are not used as they have been shown to have a higher incidence of later urethral complications.[4]

Urethral stenting is avoided and the bladder is drained with a suprapubic catheter for 3–4 weeks. Ureteral stents are left in place to overcome potential obstruction in the first 10 days following bladder closure. The pubic bones are approximated with horizontal mattress sutures of #2 nylon.

Post-operative management should include utilization of nasogastric tubes to prevent abdominal distention, appropriate pain control and limitation of movement and antimicrobial prophylaxis.[5] Following removal of traction, residual urine volume determinations are performed to ensure complete, incontinent bladder emptying. If residual volumes are increased, endoscopic evaluation of the posterior urethra may be performed to assure drainage. Renal ultrasound is performed prior to discharge and three months following surgery.

OSTEOTOMY IN PRIMARY CLOSURE

The requirement for osteotomy is determined by the extent of pubic diastasis, the size of the bladder template and the timing of the surgical procedure. Osteotomies of the pelvic ring decrease tension on the midline abdominal closure, brings the pelvic musculature together in the midline and includes the bladder neck and posterior urethra in the pelvic ring.[6] The reconstitution of the pelvic diaphragm may assist in the development of continence. Osteotomies appear to decrease the need for later urinary diversion in one published report.[7] Gearhart et al have noted a lower incidence of dehiscence and prolapse following osteotomy.[8] Pelvic fixators and Buck's traction are used for post-operative immobilization.

In a series of 40 patients referred to our institution following a failed primary closure,[9] only 6 had a prior osteotomy. The reasons for failure of the primary repair were complete dehiscence (28), bladder prolapse (10) and impassable posterior urethral strictures in 2 patients.

IMPORTANCE OF SUCCESSFUL PRIMARY CLOSURE

Successful primary closure clearly is the critical step in the development of urinary continence and spontaneous voiding. Following successful primary closure appropriate bladder growth occurs and later continence and spontaneous voiding are possible in 75–85% of patients.[1,7] If primary closure fails only 60% of patients develop an adequate capacity for later bladder neck reconstruction. When a second attempt at primary closure also fails, only 40% of patients will develop adequate bladder capacity for bladder neck reconstruction and only 20% are eventually dry. This significant reduction in functional success makes successful primary closure an important first step in exstrophy reconstruction.

CONCLUSIONS

Successful primary bladder closure is an important first step in the functional reconstruction of the patient with bladder exstrophy. Factors that contribute to a successful primary closure include appropriate timing of the procedure, good bladder and posterior urethral closure, liberal use of osteotomies, utilization of ureteral stents and bladder drainage, avoidance of urethral tubes, appropriate pain control, prevention of post-operative abdominal distention, pelvic fixation and immobilization, continued antibiotic prophylaxis and careful post-operative follow-up. With successful initial closure, the patient is well on the way to achieving maximal bladder growth and continence.

REFERENCES

1. Gearhart J.P., Jeffs R.D.: Exstrophy-epispadias complex and bladder anomalies. In Campbell's Urology 7th edition. Walsh P.C., Retik A., Stamey T., Vaughan D. eds. pp.1939–1990.
2. Gearhart J.P. Ben-Chaim J., Jeffs R.D., Sanders R.C.: Criteria for the prenatal diagnosis of classic bladder exstrophy. Ob/Gyn., 85(6) 961–964 1995.
3. Mathews R., Jeffs R.D., Taylor S., Gearhart J.P.: Combined bladder closure and epispadias repair in the management of bladder exstrophy. Submitted for publication J Urol., 1997.
4. Gearhart J.P., Peppas D.S., Jeffs R.D.: Complications of paraexstrophy skin flaps in the reconstruction of classical bladder exstrophy. J Urol., 150:627 1993.
5. Lowe F.C., Jeffs R.D.: Wound dehiscence in bladder exstrophy: An examination of the etiologies and factors for initial failure and subsequent success. J Urol., 130:312–315,1983.
6. Sponseller P.D., Gearhart J.P., Jeffs R.D.: Anterior innominate osteotomies for failure or late closure of bladder exstrophy. J Urol., 145:1151 1991.
7. Husmann D.A., McLorie G.A., Churchill B.M.: Closure of the exstrophic bladder: An evaluation of factors leading to its success and its importance on urinary continence. J Urol., 142:522 1989.
8. Gearhart J.P., Froschner D.C., Jeffs R.D., Ben-Chaim J., Sponseller P.D.: A combined vertical and horizontal pelvic osteotomy approach for primary and secondary repair of bladder exstrophy. J Urol., 155:689–693 1996.
9. Gearhart J.P., Peppas D.S., Jeffs R.D.: The failed exstrophy closure: strategy for management. Br J Urol., 71:217, 1993.

12

ANTERIOR AND POSTERIOR INNOMINATE OSTEOTOMY IN EXSTROPHY REPAIR

Paul D. Sponseller,[1] John P. Gearhart,[2] Mihir M. Jani,[1] and Robert D. Jeffs[2]

[1]Department of Orthopedics
[2]Department of Urology
The Johns Hopkins School of Medicine
Baltimore, Maryland

INTRODUCTION

Children with exstrophy are born with the bladder exposed and an open pelvic ring. Several types of pelvic osteotomies have been developed to decrease the stress on the initial exstrophy closure through closing the pelvic ring. Shultz (1958) first described the bilateral posterior iliac osteotomy as part of a two-stage repair of bladder exstrophy.[11] The earliest outcomes studies of this osteotomy were done by O'Phalen (1963) and Aadalen (1979).[8,1] This pelvic osteotomy so far has been shown to lower rates of wound dehiscence and to help obtain a more secure and better genitourinary reconstruction.[8,1]

Newer types of osteotomies that have been developed include bilateral superior pubic ramotomy,[3–5,9–10] mid-iliac diagonal osteotomy,[7] bilateral anterior innominate osteotomy,[12] and bilateral combined horizontal and vertical pelvic osteotomy.[6]

Studies of the anterior innominate osteotomy for failure or late closure of bladder exstrophy revealed several advantages over posterior iliac osteotomy. The pelvis was found to be more mobile with the anterior innominate osteotomy. In addition, the patient did not need to be turned during osteotomy and urologic repair because both were performed with the patient in the supine position.[12] The combined anterior/posterior innominate osteotomy was found to correct the pelvic anatomy more completely, and studies showed that this modified procedure of the original anterior innominate osteotomy was safe and effective in both initial and late exstrophy closure.[6]

In order to assess outcome of anterior and anterior/posterior innominate osteotomy in all stages of exstrophy repair, we analyzed all cases of these pelvic osteotomies performed at Johns Hopkins.

The Exstrophy–Epispadias Complex, edited by Gearhart and Mathews.
Kluwer Academic / Plenum Publishers, New York, 1999.

METHOD OF ANALYSIS

We reviewed all Johns Hopkins cases from 1988 to the present of children who underwent pelvic osteotomy as a part of reconstructive surgery of the bladder exstrophy complex. Thirteen children had the diagnosis of cloacal exstrophy and ninety-one the diagnosis classical exstrophy. Of these patients, 40 anterior innominate,[12] 61 combined anterior/posterior innominate,[6] 5 posterior iliac,[1,8,11] 2 suprapubic ramotomy,[3–5,9–10] and 1 posterior innominate osteotomies were performed. Five children underwent some combination of the above osteotomies at different time-points. A total of 101 anterior and anterior/posterior innominate osteotomies were performed in 98 children (mean age of 2.9 ± 3.1 years).

Anterior innominate osteotomy is performed by placing the patient in supine position and making oblique incisions inferior to the anterosuperior iliac spine as described for the Salter osteotomy and by Sponseller (1991).[12] The pelvis is subperiosteally exposed inferiorly to the pectineal tubercle and posteriorly to the sacroiliac joints. The periosteum of the sciatic notch is raised and anterior osteotomy is done by using a Gigli saw at a point halfway between the anterosuperior and anteroinferior spines. Posterior vertical osteotomy is performed by placing a small rongeur in the cancellous bone from the first osteotomy anterolateral to the beginning of the sciatic notch. The rongeur is used to create a closing wedge osteotomy vertically parallel to the sacroiliac joint.[6] The posterior cortex is kept intact. This combination of osteotomies corrects the abnormalities in both the anterior and posterior parts of the pelvis.

Two threaded fixator pins are placed in the inferior fragment and two pins are placed in ilium superiorly. Radiographs are obtained to confirm pin placement, soft tissues are closed, and the urologic procedure is performed. At the end of the procedure, the pelvis is closed using either a suture or a plate between the two pubic rami. An external fixator is then placed to hold the pelvis in corrected position. If radiographs taken 7–10 days postoperatively do not show complete reduction of symphyseal diastasis, the inferior fixator bars can be gradually approximated over several days. Patients remain supine in skin traction for approximately four weeks and are kept in external fixation for approximately six weeks.[6]

The sequence of bladder exstrophy reconstruction involved primary closure of the bladder and abdominal wall, epispadias repair when the penis was larger (around age one), and bladder neck reconstruction when the bladder was of adequate size (60 ml) and the child was behaviorally ready to maintain dryness. In a few cases bladder augmentation and stoma construction was necessary to correct for persistent incontinence or inability to reconstruct the bladder neck because of small bladder size.[2] In addition, a few children underwent sex reassignment as part of any of these stages. Patients underwent pelvic osteotomy at any of these stages depending upon previous urologic and orthopedic surgeries done before presentation. Of the children who underwent anterior innominate or anterior innominate/posterior innominate osteotomies, 33 patients had primary closure at other institutions and either had wound dehiscence and/or bladder prolapse and therefore presented for repeat primary closure. Pelvic innominate osteotomy was performed with initial exstrophy closure in 27 infants with primary presentation to Johns Hopkins. Thirty-two children underwent bladder neck plasty, one child underwent bladder augmentation, and six had miscellaneous urologic repair with anterior innominate or anterior/posterior innominate osteotomy.

Radiographic Analysis

All available radiographs were obtained and symphyseal diastasis was measured at four time points: 1) preoperatively (within two months of osteotomy), 2) postoperatively (within two weeks after osteotomy), 3) at approximately three months after osteotomy and after removal of the external fixator, and 4) at the latest follow-up. The diastasis was determined by measuring the distance between the two most medial points of the pubic rami.

Level of Continence

Children with classic exstrophy who had undergone innominate osteotomy and bladder neck plasty but not bladder augmentation and/or stoma construction were scored for continence defined as voluntary voiding through a reconstructed urethra: "excellent" — dry at day and night (sometimes with occasional stress incontinence), "satisfactory" — dry greater than 3 hours at day and wet at night, "fair" — dry 2–3 hours, and "poor" — dry less than 2 hours. The mean diastasis was calculated at the four time points as above for each group.

Age Stratification and Percent Approximation of the Pubis

Classical exstrophy patients who had undergone innominate osteotomy were stratified by age at time of osteotomy into five groups: 1) ages 0–6 months, 2) ages 6–18 months, 3) ages 18–24 months, 4) ages 2–5 years, and 5) greater than 5 years. The percent approximation of the pubis was calculated according to the following formula:

$$\frac{(\text{preoperative diastasis} - \text{postoperative diastasis})}{(\text{preoperative diastasis} - 1)} \times 100$$

Correction for the one centimeter symphyseal diastasis that is present in normal individuals is reflected in the denominator. The mean percent approximations and diastases for each time point as above were determined for each age group.

RESULTS AND DISCUSSION

Surgical Outcome and Complications

The mean estimated orthopaedic blood loss for anterior innominate and anterior/posterior innominate pelvic osteotomies was 149 ± 123.9 milliliters. Total external fixator time was 6.2 ± 1.6 weeks. For those children who had modified Bryant's traction in addition to external fixation, time in traction was 4.3 ± 1.4 weeks. For those treated with traction only, time in traction was 3.4 ± 0.4 weeks. Post-operative complications included 8 cases of transient left femoral nerve palsy which resolved in an average of 12 weeks, 3 cases of delayed union, and 19 cases of superficial pin tract infection treated with oral antibiotics.

Six cases of transient left femoral nerve palsy occurred in those who underwent anterior innominate osteotomy (five cases were sequential). Only 2 cases of transient left femoral nerve palsy occurred in those who underwent combined anterior/posterior innominate osteotomy. The frequency of left femoral nerve palsy after combined innominate osteotomy was significantly less ($p<0.05$) than in those who underwent anterior innominate osteotomy alone. There were no cases of right femoral nerve palsy.

Other complications included 1 case of superficial wound infection that required irrigation and debridement, 1 case of transient right abductor atrophy, 1 case of loosened external fixator pins, 1 left ileum infection requiring irrigation and debridement, 1 case of prominent veins on child's right side, and 1 case of transient right peroneal motor palsy. In addition, one child has a mild left Trendelenburg gait.

One child had bladder prolapse after primary closure and bilateral anterior/posterior innominate osteotomies, and two children developed wound dehiscence after primary closure and pelvic osteotomy, of which one had developed severe respiratory distress secondary to an RSV infection. These were considered the only failures resulting in a failure rate of less than 3.0% for anterior innominate and combined anterior/posterior innominate osteotomies.

Symphyseal Diastases

The mean symphyseal diastasis at each time point for classical and cloacal patients is shown in the following table:

Mean symphyseal diastases (cm):	Classical exstrophy	Cloacal exstrophy	p value
Preop	5.04±1.22	8.04±2.79	p<0.01
Postop	1.8±0.64	2.73±1.35	p<0.05
3 months F/U	2.57±0.81	4.25±1.24	p<0.01
Latest F/U	3.72±1.10	5.46±3.25	p<0.1

The cloacal mean symphyseal diastases were significantly larger than the mean classical symphyseal diastases at all time points except latest follow-up. Although children with classical exstrophy seemed to gain better correction, children with cloacal exstrophy did have some correction of intersymphyseal diastasis.

Continence and Diastasis

Of those classical exstrophy patients who had undergone osteotomy and bladder neck reconstruction but not bladder augmentation, 17 patients scored "excellent", 11 "satisfactory", 7 "fair", and 3 "poor" for degree of continence. There was no significant difference in symphyseal diastasis between any of the groups stratified for degree of continence. The sample numbers for the "satisfactory" and "poor" groups are too small to be compared to the other groups. In addition, longer follow-up is needed given the young age (< 5 years old) of most our patients and that as they behaviorally mature, they will gain more control to remain dry.

Age of Osteotomy and Percent Pubic Approximation

The percent pubic approximation for each age group at 3 months follow-up and latest follow-up is shown in the following table and Figure 1.

Percent pubic approximation

Age at surgery:	0–6 mo. (n=23)	>6–18 mo. (n=5)	>18–24 mo. (n=9)	>24–60 mo. (n=21)	>5 yrs. (n=11)
3 months F/U	42.3±39.7	67.4±14.5	62.4±14.8	66.5±15.4	66.6±27.2
Latest F/U	-8.37±48.5	17.4±14.5	40.5±17.9	43.0±20.8	47.5±29.4

At 3 months follow-up, there is significantly less pubic correction in the 0–6 months age group compared to other age groups ($p < 0.05$). However at latest follow-up, correction in the 0–6 months group is significantly less than the 6–18 months age group ($p<0.05$) and the percent correction for all other age groups ($p<0.001$), and the pubic approximation

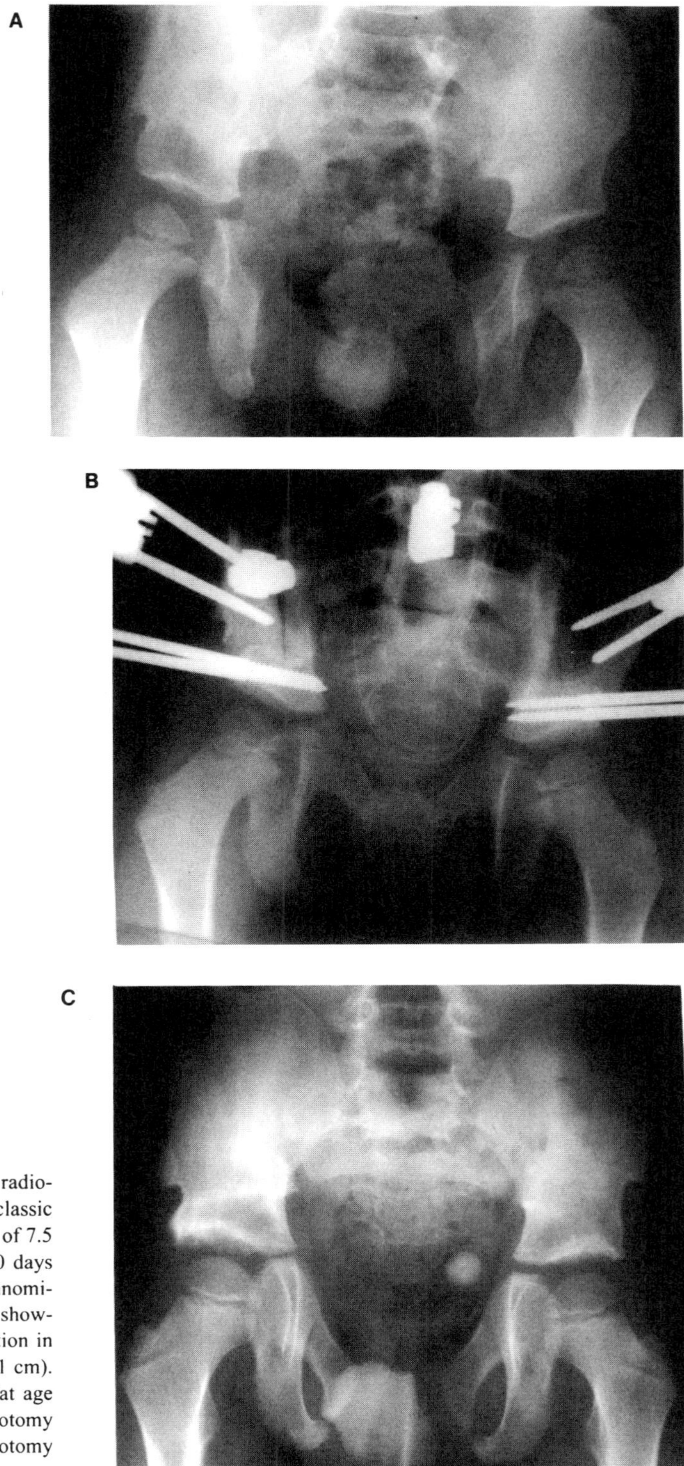

Figure 1. A: Preoperative pelvis radiograph of a four year old boy with classic exstrophy showing a pubic diastasis of 7.5 cm. B: Postoperative radiograph 10 days after combined anterior/posterior innominate osteotomy of the same patient showing fixator pin positions and reduction in diastasis to 1.1 cm (normal: 0.7 to 1 cm). C: Radiograph of patient's pelvis (at age 7) one and one half years after osteotomy showing complete healing of osteotomy sites and a diastasis of 3.1 cm.

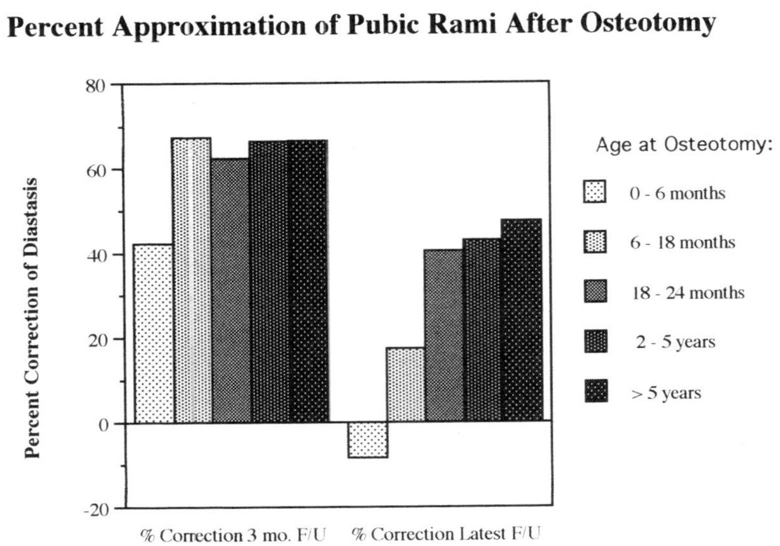

Figure 2. Graph of percent correction of intersymphyseal diastasis at 3 months and latest follow-up after anterior/posterior innominate osteotomy. At 3 months follow-up, correction for the 0–6 month group is significantly less than all other age groups. At latest follow-up, both 0–6 month and 6–18 month age groups have significantly less correction than the remaining groups.

for the 6–18 months age group is significantly less than the percent correction for the age 18–24 month ($p<0.05$), 24 month – 5 year ($p<0.01$), and the > 5 year ($p<0.02$) age groups.

The age of the patient seems to play a role in the amount of correction maintained over time. As the bone grows and ages from infancy, the cartilage of the infant skeleton ossifies. This provides a more solid bone stock for osteotomy and external fixation which allows for better maintenance of the pelvis in the corrected position.

SUMMARY

- Anterior innominate osteotomy and combined anterior/posterior innominate osteotomies are safe, efficacious procedures which can be performed as part of any stage in exstrophy repair.
    1. The failure rate is less than 3% to date.
    2. Transient left femoral nerve palsy is the only significant complication.
- Classical exstrophy patients gain more correction of symphyseal diastasis than those children with cloacal exstrophy.
- There was no difference in symphyseal diastasis among classical exstrophy patients with varying abilities to maintain dryness through spontaneous urethral voiding.
- Bone of older children maintains better correction of the repaired pelvis over time.

REFERENCES

1. Aadalen, R. J.; O'Phelan, E. H.; Chisholm, T. C.; McParland, F. A. Jr.; and Sweetser, T. H. Jr.: Exstrophy of the bladder: long term results of bilateral posterior iliac osteotomies and two-stage anatomic repair. *Clin. Orthop.*, 151: 193–200, 1980.

2. Ben-Chaim, J.; and Gearhart, J.P.: Current management of bladder exstrophy, Lesson #19. *AUA Updates*, 16: 146–151, 1997.
3. Frey, P.: Bilateral anterior pubic osteotomy in bladder exstrophy closure. *J.Urol.*, 156: 812–815, 1996.
4. Frey, P.; and Cohen, S. J.: Blasenekstrophie-anteriore Osteotomie des Beckens-eine neue operative Technik zur Erleichterung der Staabilisierung des Beckens und des Abdominalverschlusses. *Z. Kniderchir.*, 43: 171, 1988.
5. Frey, P.; and Cohen, S. J.: Anterior Pelvic Osteotomy. A new operative technique facilitating primary bladder exstrophy closure. *Brit. J. Urol.*, 64: 641, 1989.
6. Gearhart, J. P.; Forschner, D. C.; Jeffs, R. D.; Ben-Chaim, J; and Sponseller, P. D.: A combined vertical and horizontal pelvic osteotomy approach for primary and secondary repair of bladder exstrophy. *J. Urol.*, 155: 689–693, 1996.
7. McKenna, P. H.; Khoury, A. E.; McLorie, G.A.; Churchill, B. M.; Babyn, P. B.; and Wedge, J.H.: Iliac osteotomy: a model to compare the options in bladder and cloacal exstrophy reconstruction. *J. Urol.*, 151: 182–187,1994.
8. O'Phelan, E.H.: Iliac osteotomy in exstrophy of the bladder. *J. Bone and Joint Surg.*, 45-A: 1409–1422, Oct. 1963.
9. Perovic, S.; Brdar, R; and Scepanovic, D.: Bladder exstrophy and anterior pelvic osteotomy. *Brit.J. Urol.*, 70: 678–682, 1992.
10. Schmidt, A. H.; Keenen, T. L.; Tank, E. S.; Bird, C. B.; and Beals, R. K.: Pelvic osteotomy for bladder exstrophy. *J. Pediat. Orthop.*, 13: 214–219, 1993.
11. Shultz, W. G.: Plastic repair of exstrophy of bladder combined with bilateral osteotomy of ilia. *J. Urol.*,79: 453–458, 1958.
12. Sponseller, P. D.; Gearhart, J. P.; and Jeffs, R. D.: Anterior innominate osteotomies for failure or late closure of bladder exstrophy. *J. Urol.*, 146: 137–140, 1991.

13

BLADDER EXSTROPHY CLOSURE WITHOUT OSTEOTOMY IN PHILADELPHIA

Douglas A. Canning, Thomas Lanchoney, Simone McKitty, Stephen Zderic, Howard M. Snyder, III, and John W. Duckett, Jr.

The Children's Hospital of Philadelphia
Philadelphia, Pennsylvania

ABSTRACT

Purpose

To evaluate the complications following initial bladder closure and urinary continence in patients born with bladder exstrophy closed early without posterior iliac osteotomy.

Materials and Methods

A retrospective review of 105 patients treated at The Children's Hospital of Philadelphia (CHOP) since 1952. Sixty-two patients had initial treatment at CHOP. Age at initial closure, surgical procedure and complications were noted. Follow up was performed with detailed questions regarding continence. Patients completing reconstructive surgery for continence were grouped as follows: *Dryness* was recorded as dry day and night, dry day only, dry three hours or wet. *Continent* patients were those patients who had achieved at least a 3-hour dry interval without the need for intermittent catheterization. Statistical analysis was performed using Fisher's exact test.

Results

Of 62 patients who had initial therapy at CHOP, 12 had initial ureterosigmoidostomy. Initial bladder closure was performed at CHOP for the remaining 50. Of the 50, 45 had both the closure and subsequent continence procedures at CHOP. Thirty-one of 43 (71%) of patients closed without osteotomy had a successful closure. Three of 5 closed with osteotomy were successful. Only 6 (13%) of the 45 patients are voiding through the

The Exstrophy–Epispadias Complex, edited by Gearhart and Mathews.
Kluwer Academic / Plenum Publishers, New York, 1999.

native bladder with continence. Thirty (67%) of 45 are dry with intermittent catheterization. Neither age at initial closure, use of posterior iliac osteotomy nor use of paraexstrophy flaps (PEF) affected continence or complication rates of the initial bladder closure. Of only 5 patients closed with osteotomy, 2 are continent and voiding with the native bladder compared to only 4 of 20 patients closed without osteotomy (p=0.065).

Conclusions

Continence following staged reconstruction for bladder exstrophy is difficult to achieve even at a center with considerable experience. Most patients can be dry with intermittent catheterization or early closure of the bladder without osteotomy is not associated with increased risk of complications compared with closure with osteotomy. Despite few patients in this series, a trend toward improved continence in patients undergoing osteotomy was observed.

INTRODUCTION

Over 225 patients with bladder exstrophy have been treated by our team at the Division of Pediatric Urology at The Children's Hospital of Philadelphia (CHOP) since 1970. The majority of patients in our series have been closed without posterior iliac osteotomy. The use of osteotomy to facilitate the initial closure in bladder exstrophy is thought to provide additional ability to close the bladder and reapproximate the urethra and bladder neck within the levator sling to provide additional continence.[1–4] Our series, which demonstrates a commitment to early bladder closure, provides an opportunity to review the results of staged reconstruction without osteotomy.

Duckett at CHOP conceived the use of paraexstrophy flaps (PEF) to augment urethral length at the time of primary bladder closure. For most of the last 30 years, males[5] and females[6] alike have had closure with the flaps. The use of these flaps has been criticized however, with some surgeons noting a high incidence of bladder outlet obstruction associated with scarring at the bladder neck. This problem was thought due to poor blood supply at the prostatic urethra, the site of incorporation of the flap.[7] Our speculation is that well designed flaps were not hazardous and could be helpful in providing additional urethral length during the initial closure. In addition to assessing results following primary closure without osteotomy, we hoped to better define the effect of the use of paraexstrophy flaps (PEF) in the primary closure.

METHODS

We reviewed the records of 105 patients with classic bladder exstrophy treated at the Children's Hospital of Philadelphia (CHOP) between 1952 and 1996. This group represented a subset of 225 patients treated with the exstrophy variants. We excluded patients with cloacal exstrophy, epispadias and exstrophy variants. Cases in which primary ureterosigmoidostomy was performed were excluded from consideration in this analysis. Of the 105 patients, 43 patients underwent initial treatment elsewhere; leaving 62 patients with classic bladder exstrophy that had initial treatment at CHOP. These patients were primarily analyzed.

Age at initial closure was recorded, as was the type of surgical procedure. If osteotomy was used, the age at which the osteotomy was performed was recorded. Those patients who had paraexstrophy flaps at the time of the original closure or subsequently were also noted. Complications of the initial closure were recorded. Subsequent continence surgery if performed was recorded as well.

All follow up visits were performed with routine physical examinations and detailed questions regarding continence. Those patients who had completed reconstructive surgery for continence were grouped as follows: *Dryness* was recorded as dry day and night, dry day only, dry three hours or wet. *Continent* patients were defined as those patients who had achieved at least a three-hour dry interval without the need for intermittent catheterization.

Statistical analysis was performed using Fisher's exact test. Statistical significance was considered when the p value was less than 0.05.

RESULTS

Although most patients had bladder closure as the initial treatment, a number of patients treated early in the series had ureterosigmoidostomy, usually at age one. Of the 62 patients who had initial treatment at CHOP, 12 patients treated early in the series had primary ureterosigmoidostomy. Initial bladder closure was performed at CHOP in the other 50 cases. The child's age at initial surgery ranged from 24 hours to 22.3 years. Follow up time ranged from 11 months to 49.6 years. Of the 50 children with the primary bladder closure at CHOP, 45 patients had both their initial bladder closure and continence procedures at CHOP. Only these patients were assessed for continence. Twelve of these patients had subsequent ureterosigmoidostomy at some point in their care. These patients were not considered in the analysis of continence as a function of osteotomy, timing of bladder closure, or use of paraexstrophy flaps.

True continence (a dry, voiding patient) was difficult to achieve. Only 6 patients of the 45 (13%) are continent as defined as voiding and dry. Of these 6, 5 are dry during the day and night. One patient is dry for intervals of at least 3 hours. Of the six continent patients, 5 had a Young-Dees-Leadbetter (YDL) bladder neck plasty procedure[8] and one had a Koff cinch.[9] Dryness in patients emptying the bladder with intermittent catheterization was much easier to achieve. Thirty patients are dry but not continent (emptying with intermittent catheterization). Of these, 26 are dry both day and night, 3 are dry during the day and 1 is dry for 3-hour intervals. Three of the 45 patients are wet and awaiting further surgery. Six patients are lost to follow up.

Closure without Osteotomy

A large majority of our patients were closed early without osteotomy. Forty-three (86%) of the 50 primary bladder closures at CHOP were closed without osteotomy. In most cases, these patients were closed within 72 hours of birth. Five patients who were closed after 72 hours of age were closed with osteotomy. Osteotomy status was unknown in two patients closed early in the series.

Complications of the Initial Closure. There was no statistical difference in complication rate between those closed with and without osteotomy. Thirty-one of the 43 patients (71%) closed without osteotomy had a successful bladder closure. Of the 12 patients with complications, 3 suffered dehiscence, 7 developed bladder outlet obstruction and 2 developed bladder stones. Two of the 5 patients closed with osteotomy suffered complications. One patient developed bladder outlet obstruction. Another had a dehiscence.

Continence and Dryness. Patients who had osteotomy as part of the closure seemed to do better achieving continence following continence surgery. Initial closure and continence surgery on the native bladder was performed at CHOP in 23 patients exclusive of those who subsequently had ureterosigmoidostomy. Of these children, 20 had bladder closure without osteotomy. Two patients had osteotomy done at the time of the initial closure. Both patients who had osteotomy were continent while only 4 of 20 patients without osteotomy were continent ($p=0.065$). There were nine patients who had bladder closure without osteotomy and were closed later than at 48 hours of age. Of these, two were continent. There was no difference in continence in this group compared with those undergoing early closure without osteotomy.

Dryness was much easier to achieve. Virtually all patients regardless of osteotomy status were dry following continence surgery. Most of these patients, however, required augmentation cystoplasty to provide sufficient bladder volume.

Closure with Paraexstrophy Flaps

Forty-three (86%) of the 50 children undergoing primary closure at CHOP were closed using paraexstrophy flaps. The other seven patients were closed without the use of flaps.

Complications of the Initial Bladder Closure. Complications following the initial bladder closure were similar whether paraexstrophy flaps were or were not used. Of the 43 patients closed with PEF there were no complications in 27 (63%). There were 5 patients who suffered a dehiscence and 7 patients that had bladder outlet obstruction. Four patients developed bladder stones, which may have been due to poor bladder emptying. Of the 7 closed without PEF, 5 (71%) had no complications from the initial closure, one had bladder outlet obstruction and one developed bladder stones.

Continence Following Closure with PEF. There was no difference in continence or dryness whether or not PEF was used in the initial bladder closure. Of the 45 patients who had closure and continence procedures at CHOP (excluding those patients with ureterosigmoidostomy), 4 patients had been closed without PEF, 19 had closure with PEF and records were unclear in 2. Similar continence rates for the voiding group and similar dry intervals for those patients who empty with a catheter were noted. All 4 patients without PEF are dry but not continent. Of 19 patients with PEF, 4 are continent and 13 are dry. Two patients are wet and await further surgery.

DISCUSSION

Despite the limitations of this retrospective review, several findings seem clear. First, despite our ability to make nearly all children born with bladder exstrophy dry, relatively few in most major series are truly continent. Our group, with only 13% of all patients voiding with day and night continence, although disappointing, seems comparable to that of others.[10–12] Both the surgeon and the parent feel tremendous excitement when a child voids to completion following a successful reconstruction. We must be careful to control enthusiasm when we quote success rates to parents. We must be careful to distinguish between continence (a dry voiding patient) from dryness (a dry patient who has to empty the bladder with a catheter) and be cognizant of the potential limitations of the bladder in children born with exstrophy.[13,14]

As in many retrospective series, comparing dissimilar sized groups can be difficult. At CHOP during this period, few patients who were closed early had osteotomy performed. With the small numbers in the osteotomy group, comparisons are difficult. Incidence of bladder outlet obstruction and dehiscence rate was similar in both groups. Our series shows that at least for the first stage of the reconstruction for classic bladder exstrophy, little harm comes from avoiding osteotomy if the closure is done early. These observations have been noted in other large series of patients where closure with osteotomy was more common.[2,15]

Despite the small number of our patients who had osteotomy as part of the initial closure, 2 of our 6 continent patients had osteotomy. We think that statistical significance would probably have been evident had we had more patients in the osteotomy group. The observation that continence is easier to achieve in those patients with osteotomy has been noted by others but is notoriously difficult to prove in retrospective studies.[2,15] Despite no statistical evidence that osteotomy provides a safer initial closure, we believe that continence may be easier to achieve in those patients who have osteotomy as part of the initial closure. We are now more aggressive about recommending osteotomy in a staged or single stage reconstruction.

Despite the similar complication rates for the initial closure with and without paraexstrophy flaps, we now rarely use them. Jeff's technique of notching the lateral bladder neck tissue in a series of "Z" plasties is effective and can be incorporated into the staged repair without disruption of the bladder neck or prostatic urethra.[16] Mitchell's repair,[17] which takes advantage of wide exposure provided by splitting the corporal bodies as the bladder neck is closed seems to provide the additional urethral length to provide for placement of the bladder neck within the levator sling. Kelley's closure, which stages the repair while also providing for radical mobilization of the bladder neck at the level of the levator muscle group may also be a good alternative, especially if a shorter duration surgery is desired in the newborn.[18]

We wish to thank our friend, John Duckett for his leadership, enthusiasm and compassionate commitment to pediatric urologic surgery. John's thoughtful and innovative approaches to this difficult problem have inspired all of us who care for children born with bladder exstrophy.

REFERENCES

1. Gearhart, J.P., *et al., The multiple reoperative bladder exstrophy closure: what affects the potential of the bladder?* Urology, 1996. **47**(2): p. 240–3.
2. Husmann, D.A., G.A. McLorie, and B.M. Churchill, *Closure of the exstrophic bladder: an evaluation of the factors leading to its success and its importance on urinary continence.* J Urol, 1989. **142**(2 Pt 2): p. 522–4.
3. Lottmann, H.B., *et al., Bladder exstrophy: evaluation of factors leading to continence with spontaneous voiding after staged reconstruction.* Journal of Urology, 1997. **158**(3 Pt 2): p. 1041–4.
4. Jeffs, R.D., *Functional closure of bladder exstrophy.* Birth Defects, 1977. **13**(5): p. 171–3.
5. Duckett, J.W., *Use of paraexstrophy skin pedicle grafts for correction of exstrophy and epispadias repair.* Birth Defects, 1977. **13**(5): p. 175–9.

6. Spindel, M.R., B.H. Winslow, and G.H. Jordan, *The use of paraexstrophy flaps for urethral construction in neonatal girls with classical exstrophy.* J Urol, 1988. **140**(3): p. 574–6.
7. Gearhart, J.P., D.S. Peppas, and R.D. Jeffs, *Complications of paraexstrophy skin flaps in the reconstruction of classical bladder exstrophy.* Journal of Urology, 1993. **150**(2 Pt 2): p. 627–30.
8. Leadbetter, G.W.J., *Surgical correction of total urinary incontinence.* J. Urol., 1964. **7**: p. 1–32.
9. Koff, S.A., *A technique for bladder neck reconstruction in exstrophy: the cinch.* J Urol, 1990. **144**(2 Pt 2): p. 546–9.
10. Mollard, P., P.D. Mouriquand, and X. Buttin, *Urinary continence after reconstruction of classical bladder exstrophy (73 cases).* Br J Urol, 1994. **73**(3): p. 298–302.
11. Connor, J.P., *et al.*, *Long-term followup of 207 patients with bladder exstrophy: an evolution in treatment.* J Urol, 1989. **142**(3): p. 793–5.
12. de la Hunt, M.N. and B. O'Donnell, *Current management of bladder exstrophy: a BAPS collective review from eight centres of 81 patients born between 1975 and 1985.* J Pediatr Surg, 1989. **24**(6): p. 584–5.
13. Canning, D.A., *Bladder exstrophy: the case for primary bladder reconstruction [editorial].* Urology, 1996. **48**(6): p. 831–4.
14. Hohenfellner, R. and R. Stein, *Primary urinary diversion in patients with bladder exstrophy [editorial].* Urology, 1996. **48**(6): p. 828–30.
15. Canning, D., *et al.*, *A computerized review of exstrophy patients managed during the past thirteen years.* J. Urol., 1989. **141**: p. 224A.
16. Gearhart, J. and R. Jeffs, *Exstrophy-Epispadias complex and bladder anomalies*, in *Campell's Urology*, P. Walsh, *et al.*, Editors. 1998, WB Saunders: Philadelphia. p. 1939–1990.
17. Ngan, J., *et al.*, *Factors contributing to urinary continence in classic exstrophy.* J. Urol, 1996. **155**(5): p. 483A.
18. Kelley, J., *Vesical exstrophy: repair using radical mobilisation of soft tissues.* Pediatric Surgery International, 1995. **10**: p. 298–304.

14

BILATERAL ANTERIOR PUBIC OSTEOTOMY IN BLADDER EXSTROPHY CLOSURE

Peter Frey

Department of Pediatric Surgery, Centre Hospitalier
Universitaire Vaudois, Lausanne, Switzerland

ABSTRACT

Anterior pelvic osteotomy of the superior ramus of the pubic bone is a simple and efficient method to facilitate symphyseal approximation and abdominal wall closure without or with low tension on the suture lines in neonates operated for bladder exstrophy. Older children for whom operation has been delayed can also benefit from this method. Compared with all other osteotomies in exstrophy surgery it does not require additional incisions nor does the patient need to be repositioned on the operating table. Successful operation does not depend on any particular orthopedic skills and can easily be done by the pediatric urologist. We report our clinical experience with this alternative technique which we have performed in 16 patients who have been operated for bladder exstrophy. In particular the technique and its difficulties are discussed. The immediate postoperative results regarding abdominal wall closure were excellent in all 16 patients. Major postoperative complications however, were seen in two patients. Despite antibiotic prophylaxis, one child developed a severe soft tissue infection resulting in complete bladder dehiscence. In a further patient, an obturator nerve injury, resulting in transient palsy which recovered completely, was seen. The first complication was not related to the osteotomy; the second however, definitely was.

INTRODUCTION

Classical bladder exstrophy, including total epispadias, is a rare complex congenital malformation occurring in one in 10,000 to one in 50,000 live births with a male predominance of 2–3:1.[1] It represents a multi-organ malformation commonly involving the genitourinary tract and the musculoskeletal system. The extreme form—the cloacal exstrophy—in which, the intestinal tract is also malformed, is observed in an equal male-female ratio with a prevalence of 1:200,000 live births.

The Exstrophy–Epispadias Complex, edited by Gearhart and Mathews.
Kluwer Academic / Plenum Publishers, New York, 1999.

Common to all different degrees of exstrophy is the failure of the fusion of the midline abdominal tissue in the embryo,[2] due to inhibition of mesodermal ingrowth by lack of the regression of the cloacal membrane.[1] This failure results—in classic exstrophy—in a defect of the lower abdominal wall represented by rectus and symphyseal diastasis, dysmorphic pelvis, producing anteriorly open respectively non-fused bladder and non-tubularized epispadiac urethra. In cloacal exstrophy, the exstrophic bladder is, furthermore, separated by the prolapsing intestine.

The primary aim of any exstrophy surgery is to reconstruct an adequate and functional bladder and posterior urethra. Epispadias repair is normally done as a second stage procedure at approximately one year of age. It is of functional importance that the reconstructed bladder is transposed into an intra-abdominal position and that a near to normal retro-symphyseal angle of the posterior urethra is established.

To achieve this goal, complete and utmost tension free abdominal wall closure and adequate symphyseal approximation of the previously widely separated pubic bones are mandatory. Unless this approximation is performed within 48 hours of birth, and even then, some form of osteotomy might become necessary.[3,4] In the past, several orthopedic techniques of osteotomy have been described.

Bilateral iliac osteotomy is the method most commonly applied so far.[2,3,5,6] Although it facilitates re-approximation of the symphysis and subsequent abdominal wall closure, it does represent an extensive procedure necessitating orthopedic skills, and is not free of complications. In addition, the patient needs to be repositioned for urological surgery, the procedures requires two further incisions and the duration of the operation and of anesthesia are considerably prolonged. Anterior, iliac or innominate osteotomy was recently described as an efficient alternative method offering adequate pelvic mobilization for excellent symphyseal approximation, which, additionally did not require the patient to be turned.[7,8] Through an anterior iliofemoral approach to the pelvis, the ilium is osteotomized horizontally or slightly diagonally using a Gigli saw or bone cutting forceps in younger children.

Combined bilateral osteotomy of the superior and inferior ramus of the pubic bone has also been reported as a successful technique for achieving symphyseal approximation.[9,10] The disadvantages of this method are the additional two incisions and the undermining of the skin for a relatively long distance in order to get satisfactory access to the inferior rami. Although this operation does not require the patient to be turned, it does require orthopedic skills and it also lengthens operating time.

In an attempt to simplify the technique of osteotomy, we have developed the bilateral osteotomy of the superior ramus of the pubic bone alone—a procedure which can easily be performed by the pediatric urologist and which only slightly increases overall operating time.

We first reported our limited but encouraging experience with this technique in the late eighties.[11,12] In the early nineties Perovic et al.[8] and Schmidt et al.[2] also reported very promising results having applied the same technique. Perovic demonstrated the suitability of this technique in even older children, for whom operation had been delayed or who needed secondary operation for failed primary closure.

The aim of this review is to evaluate the results we have obtained since we first described the technique of bilateral superior pubic ramotomy in exstrophy repair.

MATERIAL AND METHOD

In addition to the previously reported four cases[11,12]—included in this evaluation—we have performed this technique on a further 12 patients. The latter have been operated over the last five years. From the total of 16 patients 10 were male and 6 female. Nine patients were of Caucasian and 7 were of African origin. One of these patients had previously suffered failed exstrophy closure abroad. The age of the children varied from 2 days to 3.3 years. Patients issuing from the local population were normally operated between the third and seventh day after birth. We have now completely abandoned surgery within the first 48 hours of life. Patients of African origin have almost always undergone late surgery as they had to be diagnosed first and then transported to our center by a charity organization.

Once general anesthesia is established, the edges of the ectopic bladder are freed from the abdominal wall, the bladder is closed and the posterior urethra reconstructed according to Jeffs.[5,13] When urogenital reconstruction is completed, the abdominal skin and its subcutaneous tissues are widely mobilized bilaterally. The rectus muscle and its insertion into the pubic bone is located and its sheath is opened at the lateral border. The superior ramus of the pubic bone is easily approached through this window. The insertions of the gracilis and the three adductor muscles are identified. The superior ramus is then demonstrated between the adductor muscles and the pectineus muscle, which needs to be partially dissected free from its insertion. The periosteum of the superior ramus is incised at its superior hemi-circumference medial to the partially free pectineus muscle and medial to the insertion of the inguinal ligament. The latter occasionally needs to be freed and transposed laterally to allow adequate access to the superior ramus. The incised periosteum is freed from the bone on each side taking care not to rupture its inferior continuity (Figure 1A).

Osteotomy is then performed with a chisel, being particularly careful not to tear the inferior periosteum and not to injure the obturator nerve. It is advisable to introduce a Hohmann bone retractor into the obturator foramen underneath the superior ramus in order to optimally protect nerve and vessels (Fig. 1B). The medial aspect of both superior rami are tilted towards the midline and are approximated with little or no tension using a few resorbable sutures placed deep into the cartilaginous symphysis, therefore stabilizing the anteriorly closed pelvic ring (Fig. 1C). In older children the pelvic ring can be furthermore stabilized by an external fixater.

At this stage of the procedure care has to be taken not to interpose the reconstructed posterior urethra in between the approximated symphysis. After placing two wound drains over each osteotomy, the rectus muscles and their sheaths are easily joined and are sutured together using multiple resorbable sutures. The previously created windows lateral to the rectus sheaths are also closed with interrupted resorbable sutures. The generously mobilized skin flaps are normally approached without any tension and are sutured together with fine non-resorbable sutures. The umbilicus is either reconstructed, or, if ever possible, the primary one is preserved but transposed cranially. Prophylactic large spectrum antibiotics are administered for two weeks. The upper legs are immobilized with a mermaid bandage for two to three weeks postoperatively.

RESULTS

Bladder reconstruction was technically very difficult in one patient due to an underdeveloped exstrophic bladder. Approximation of the symphysis, the rectus muscles and sheaths, as well as the abdominal skin was easily performed in all patients without or with low tension on the suture lines. Figure 2 shows the radiological appearance of the superior ramus of the pubic bone before and after osteotomy. Figure 3 demonstrates the osteotomy on spiral computer tomography with three-dimensional reconstruction. Extensive post-op-

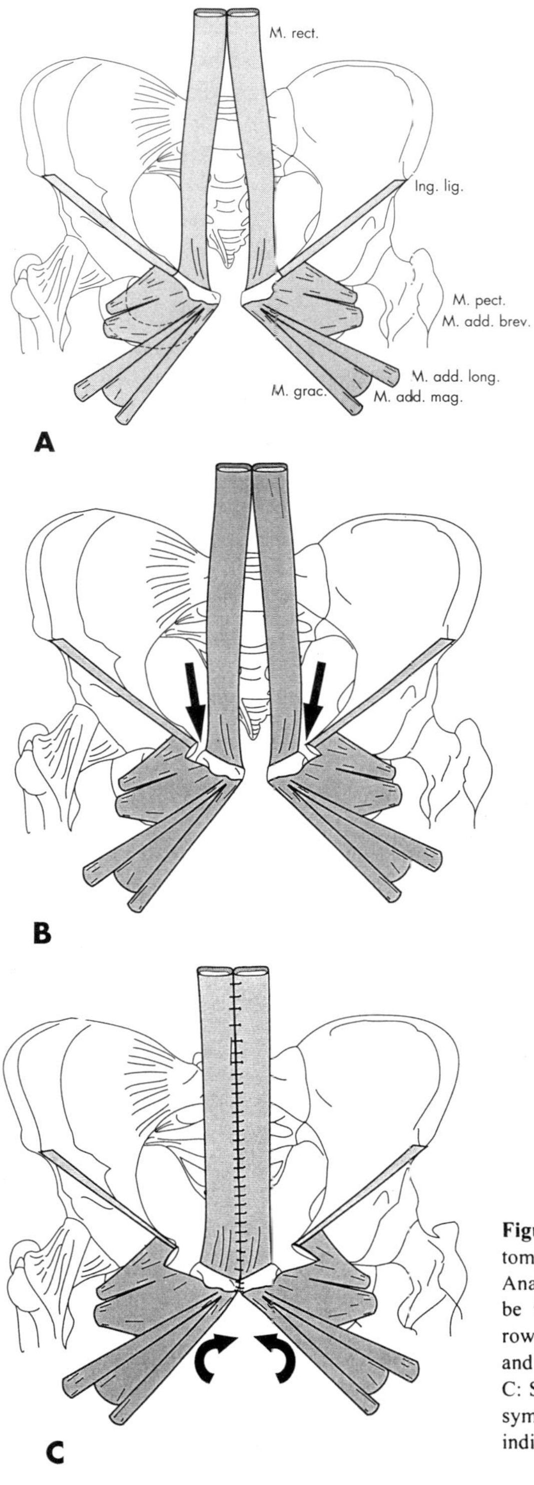

Figure 1. Schematic demonstration of anterior osteotomy of the superior ramus of the pubic bones. A: Anatomy of the pelvis and the muscles which have to be identified prior to osteotomy. B: Osteotomy (arrows) between adductor muscles and pectineus muscle and the insertion of the inguinal ligament respectively. C: State after performing bilateral osteotomy and after symphyseal and rectus muscle approximation. Arrows indicate tilting towards the midline.

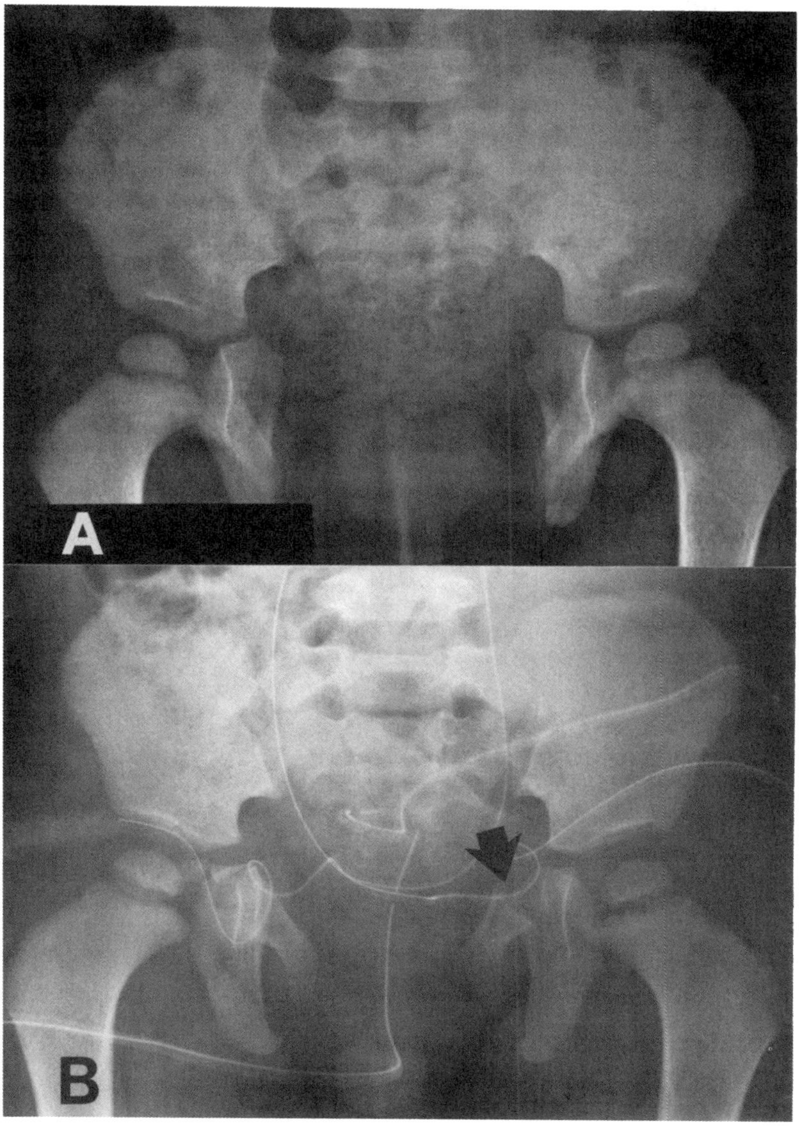

Figure 2. Radiological appearance of the pelvic. A: Before, and B: After osteotomy (arrow) of the superior ramus of the pubic bones.

erative bleeding was never observed. so that the bilaterally placed wound drains could always be removed 24 to maximum 48 hours postoperatively.

Minor wound dehiscences, closing spontaneously, were observed in two children; in one of them with a very small bladder wall prolapse, not requiring any treatment. A further patient, however, showed complete breakdown of the repair, subsequent to a severe deep wound infection. A minor superficial wound infection—healing spontaneously—was observed in two patients.

In no case was osteitis of the osteotomized pubic bones noticed. In four patients urine candida infection was found; all of them were self-limiting and no signs of systemic fungal infection were ever present.

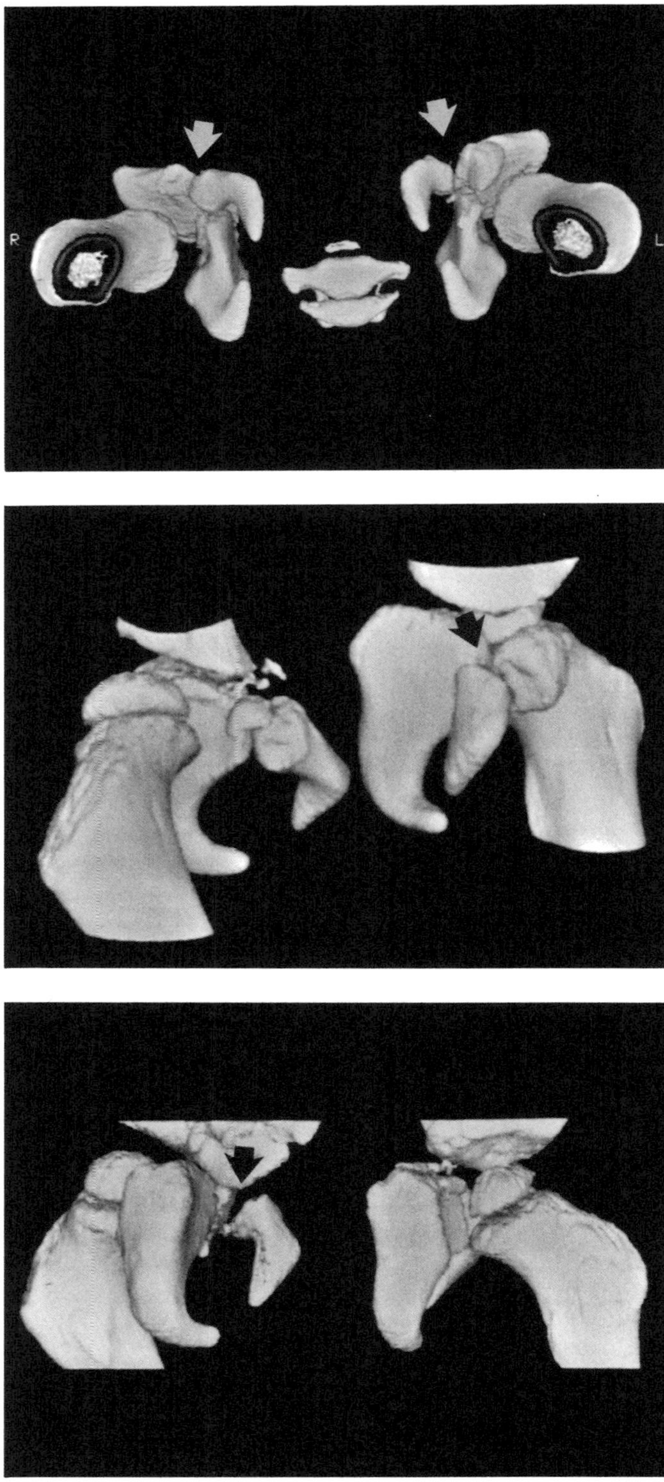

Figure 3. Anterior osteotomy (arrow) of the superior ramus of the pubis demonstrated on spiral computer tomography with three-dimensional reconstruction. A: View of the horizontal plane, B: Oblique anterio-posterior view, C: Oblique posterio-anterior view.

Transient unilateral partial adductor nerve palsy, presenting with an adductor muscle weakness was observed in one child. The same patient also suffered from a transient unilateral weakness of the perineus muscle. One three year old boy developed a postoperative left indirect inguinal hernia. Hip subluxation or dislocation were not seen in our series.

DISCUSSION

Bilateral posterior iliac osteotomy is a recognized and well established method to support reconstruction of the lower urinary tract and to facilitate symphyseal approximation and abdominal wall closure in exstrophy patients. It is, however, a time consuming major operation commonly performed by the orthopedic surgeon only. Furthermore, the procedure requires the repositioning of the patient for the urological part of the operation, which considerably increases the duration of the operation and of anesthesia. Despite the fact that bilateral posterior iliac osteotomy may initially correct pathological hip alignment, excessive external rotation of the lower extremities and the resulting waddling gait, little is known about its long-term orthopedic benefit. There is evidence that orthopedic abnormalities—typical for exstrophy patients—are not necessarily corrected by orthopedic procedures such as the bilateral posterior iliac osteotomy[14] and that, in non-operated patients, they also resolve spontaneously by adulthood.[2]

Bilateral anterior innominate osteotomy of the ileum has recently been suggested as the method of choice to ease exstrophy surgery.[7,15]

It is no doubt an efficient procedure without the drawbacks of posterior osteotomy. However, it remains a major procedure reserved for the orthopedic surgeon. Subsequent to the experience of Cook et al.,[10] and Edgerton et al.,[9] our aim was to simplify the procedure as such, so that it can easily be performed by the pediatric urologist. To our astonishment, osteotomy of the superior pubic ramus alone, always allowed approximation of the widely separated symphysis, and reconstruction of the abdominal wall with low tension or with no tension whatsoever in the newborn period. A symphyseal gap measuring 3 to 4 cm can easily be bridged as the elastic cartilaginous ischio-pubic junction acts as a sort of articulation. Encouraged by our preliminary results,[11,12] we applied the superior ramotomy of the pubic bone in older children with previously failed or delayed operations. In accordance with Perovic et al.,[8] we also achieved very successful results regardless of the age of the child. We entirely agree with the suggestion of Perovic that, once the ischio-pubic junction has ossified and lost its elasticity, the inferior ramus has to be fractured to gain sufficient mobility of the symphysis for its approximation. The anterior osteotomy of the superior ramus of the pubic bone fulfills the commonly demanded criteria for successful reconstruction of bladder and urethra, tension-free symphyseal approximation and abdominal wall closure. According to the literature,[2,7,8] any osteotomy seems to prevent dehiscence of the reconstructed bladder and abdominal wall. In addition, the levators and puborectalis sling are approximated; the pelvic floor is reconstructed. The bladder neck and the urethra are included within the pelvic ring. The continence rate might be improved.[7] Furthermore, osteotomy can increase penile length by approximation of the corpora.[9,16] The effect of superior ramotomy on penile lengthening, however, seems to be limited; this can be explained by the fact that the part of the bone where the corpora insert is only marginally transposed medially.

In our series, bladder reconstruction, even in the re-operated patient, was always fairly easy, apart from the case of one newborn, who presented with a very small dysplastic exstrophic bladder plate. Once the plate had been successfully been closed, the bladder volume was measured to be only approximately 5 mls. This bladder, however, increased its volume gradually thereafter. Bilateral anterior osteotomy of the superior pubic ramus allowed symphyseal approximation and abdominal wall closure in all patients, even in those who were older or undergoing re-operation. In accordance with Perovic,[8] we often noticed a re-diastasis of the approximated symphysis on follow-up examination, a fact which is of no clinical importance. The unique aim of the anterior osteotomy of the superior pubic ramus is to allow optimal reconstruction of the bladder, the posterior urethra, the pelvic floor and the abdominal wall regardless of possible secondary re-diastasis of the symphysis. In this context, should be mentioned that we completely omitted the originally suggested effort to preserve the effect of the osteotomy by interposing a rib transplant.[11,12]

The development of minor wound dehiscence observed in two children can be explained by local under-perfusion of skin areas subsequent to too generous a mobilization of the abdominal wall skin. These wound dehiscences always healed spontaneously. The partial bladder prolapse observed in one patient in our early series was definitely of iatrogenic origin and was due to the reconstruction of a posterior urethra, which was far too large. The latter was refashioned at the time of epispadias repair approximately one year after primary reconstruction. We cannot explain the development of a severe deep wound infection in one child, resulting in subsequent complete dehiscence of the repair, despite the administration of broad spectrum antibiotics. One can only speculate that following under-perfusion of the tissues, due to excessive devascularization during operation, the antibiotics could not reach adequate tissue concentrations to develop their full activity. The generally administered prophylactic antibiotics are not only indicated for the prevention of soft tissue infection but also to prevent the feared complication of osteitis, which we fortunately never experienced. The repeatedly observed localized candida urine infection was most probably due to the multiple indwelling catheters and cleared up spontaneously shortly after their removal. The most severe complication—directly related to surgery—was transient obturator nerve palsy, seen in one child, which recovered completely. It was most likely due to a blunt trauma by the Hohmann bone retractor introduced for protection of the nerve and vascular structures into the obturator foramen during osteotomy! The additional transient injury of the perineus nerve in the same patient was assumed to be due to a too tight mermaid bandage having been applied to the upper legs to prevent abduction of the lower extremities representing a danger which might compromise the complex repair.

Regarding immobilization of the hip and the osteotomized pelvis, we became very much in favor of the method, practiced by Gibbon et al.,[15] maintaining the reconstructed pelvic ring in a position with the legs in abduction and internal rotation. We consider to introduce this method of fixation in the future. Contrary to these authors, we think that a fixation of three to a maximum of four weeks is sufficient.

In contrast to Perovic's statement[8] that hernia formation was the main disadvantage of this method, herniotomy became only necessary in one patient, aged three years, in our series. It is quite possible that the risk of hernia formation increases in relation to age, which would explain Perovic's results in his series of older children. Using anterior osteotomy, the exstrophy closure does not need to be done within the first two days of life. The parents and especially the mother recovering from birth have therefore a chance to get accustomed to the severity of the pathology and are subsequently far better prepared to support their child psychologically during the difficult and lengthy period of repeated operations and hospitalizations. The anterior osteotomy of the superior ramus of the pubic bone seems to be an excellent method to facilitate bladder exstrophy repair and increases the duration of the operation and anesthesia. No further skin incisions, nor turning of the patients are necessary and, most importantly, it can be performed by the pediatric urologist who is generally not in command of the specific orthopedic techniques required for sophisticated osteotomies.

REFERENCES

1. Sponseller PD, Bisson LJ, Gearhart JP, Jeffs RD, Magid D, Fishman E: The anatomy of the pelvis in the exstrophy complex. J Bone Joint Surg Am 77:177, 1995.
2. Schmidt AH, Keenen TL, Tank ES, Bird Ch B, Beals RK: Pelvic osteotomy for bladder exstrophy. J Pediatr Orthop 13:214, 1993.
3. Allen TD, Husmann DA, Bucholz RW: Exstrophy of the bladder: Primary closure after iliac osteotomies without external or internal fixation. J Urol 147:438, 1992.
4. Ansell JS: Surgical treatment of exstrophy of the bladder with emphasis on neonatal primary closure: Personal experience with 28 consecutive cases treated at the University of Washington Hospitals from 1962 to 1977: Techniques and results. J Urol 121:650, 1979.
5. Jeffs RD, Guice SI, Oesch I: The factors in successful exstrophy closure. J Urol 127:974, 1982.
6. McKenna PH, Khoury AE, McLorie GA, Churchill BM, Babyn PB, Edge JH: Ilial osteotomy: A model to compare the options in bladder and cloacal exstrophy reconstruction. J Urol 151:182, 1994.
7. Sponseller PD, Bisson LJ, Gearhart JP, Jeffs RD: Anterior innominate osteotomies for failure or late closure of bladder exstrophy. J Urol 146:137, 1991.
8. Perovic S, Brdar R, Scepanovic D: Bladder exstrophy and anterior pelvic osteotomy. Br J Urol 70:678, 1992.
9. Edgerton MT, Gillenwater JY: A new surgical technique for phalloplasty in patients with exstrophy of the bladder. Plast Reconstr Surg 78:399, 1986.
10. Cook FE Jr, Leslie JT, Brannon EW: Preliminary report: A new concept of abdominal closure in infants with exstrophy of the bladder. J Urol 87:823, 1962.
11. Frey P, Cohen SJ: Blasenekstrophie - anteriore Osteotomie des Beckens - eine neue operative Technik zur Erleichterung der Stabilisierung des Beckens und des Abdominalverschluses. Z Kinderchir 43:171, 1988.
12. Frey P, Cohen SJ: Anterior pelvic osteotomy. A new operative technique facilitating primary bladder exstrophy closure. Br J Urol 64:641, 1989.
13. Jeffs RD: Exstrophy, epispadias, and cloacal and urogenital sinus abnormalities. Pediatr Clin North Am 34:1233, 1987.
14. Sutherland D, Pike L, Kaufman K, Mowery C, Kaplan G, Romanus B: Hip function and gait in patients treated for bladder exstrophy. J Pediatr Orthop 14:709, 1994.
15. Gibbon AJ, Maffulli N, Fixsen JA: Horizontal pelvic osteotomies for bladder exstrophy. A preliminary report. J Bone Joint Surg Br 73:896, 1991.
16. McLorie GA, Salter RB: Penile deformity in bladder exstrophy: Correlation with closure of pelvic defect. J Pediatr Surg 16:201, 1991.

15

THE FAILED EXSTROPHY CLOSURE

Strategies for Management—the Johns Hopkins Experience

John P. Gearhart*and Ranjiv Mathews

Johns Hopkins Hospital and School of Medicine
James Buchanan Brady Urological Institute
Division of Pediatric Urology
Baltimore, Maryland

INTRODUCTION

Secure initial bladder closure is a critical first step in the reconstruction of the patient with bladder exstrophy. The potential for gain in bladder capacity to allow later bladder neck reconstruction (BNR) appears to be dependent on success of initial closure. Increases in bladder capacity occurred more rapidly and higher continence rates (83%) correlated with successful primary closure.[1] Once primary repair fails the potential for continence drops markedly. With more than two failed closures only 23% of patients gain adequate capacity for BNR and only 50% of these attain continence.[2]

Failure of primary closure can present as dehiscence of the abdominal closure and prolapse of the bladder, prolapse of the bladder through a large urethral meatus or posterior urethral obstruction secondary to a posterior urethral stricture in patients closed with paraexstrophy skin flaps.

FACTORS CONTRIBUTING TO FAILED CLOSURE

Multiple factors appear to play a role in leading to failure of primary closure. The incidence of failed closure was higher in patients who did not have osteotomy.[3] Abdominal distension from post-operative bowel hypo-mobility and loss of urinary drainage catheters leading to bladder distension and subsequent prolapse also have been identified

* Contact author: John P. Gearhart MD, Professor and Director, Division of Pediatric Urology, Brady Urological Institute, 600 N. Wolfe Street, Baltimore, MD 21287-2101

The Exstrophy–Epispadias Complex, edited by Gearhart and Mathews.
Kluwer Academic / Plenum Publishers, New York, 1999.

as a potential causes. Stimulation of the bladder by the presence of urethral tubes has been recognized as a factor leading to spasm and further stress on the abdominal wall closure.[2] Infection leading to wound dehiscence can lead to failure of primary closure. Peri-operative antibiotic prophylaxis should be used routinely in all patients undergoing primary closure. Reconstruction of the bladder outlet should allow adequate resistance to stimulate bladder growth while preventing obstruction.

ROLE OF OSTEOTOMY IN PREVENTING FAILURE OF INITIAL CLOSURE

In a series of patients treated in Toronto, the incidence of failure was higher in patients who did not have osteotomy.[3] Seventy five percent of patients treated without an osteotomy failed compared to 35% who had undergone initial osteotomy. Prevention of patient mobility and appropriate traction devices are critical in the immediate post-operative period to prevent wound mobility and breakdown. The current use of pelvic fixators has further helped to improve nursing care and the results of osteotomy.

INITIAL OPTIONS FOR MANAGEMENT OF FAILED CLOSURE

Once prolapse or complete dehiscence of the repair is noted, initial care should focus on appropriate local measures to protect the fragile bladder mucosa. A second attempt at bladder closure should be delayed for 4–6 months. Early reclosure tends to result in repeated failure and further scarring, worsening the potential for later continence.[4] Urinary diversion and bladder closure (in 4–6 months) may also be considered. Patients with a very small bladder plate and minimal potential for normal bladder capacity may be considered for bladder excision at the time of repeat abdominal wall closure.[5,6] Alternatively reclosure and bladder augmentation combined with an outlet procedure and continent stoma may be performed. In patients presenting with outlet obstruction as the cause of failure, the upper tracts are at considerable risk. Outlet obstruction should be managed with dilatation or revision.

TIMING OF RECLOSURE

Timing of reclosure is critical for eventual successful functional outcome. Too early closure (see above) or too late closure are both associated with poor outcomes. Gearhart et al reported on 23 patients referred following multiple failed closures.[6] Fourteen had reclosure performed within 6 weeks of initial failed closure and 6 were reclosed more than two years following initial repair. None of these 20 patients achieved adequate capacity for bladder neck reconstruction. Histology demonstrated chronic inflammation and scars with increases in type III collagen. The authors recommend waiting a minimum of 4 months prior to attempting reclosure. Reclosure with osteotomy may be attempted at the earliest opportunity following this period of waiting.

SURGICAL FEATURES IN THE RECLOSURE OF FAILED EXSTROPHY

Osteotomy

Osteotomy is an integral part of the reclosure of the failed bladder exstrophy patient at the Johns Hopkins Children's Center. The pelvis in the patient with exstrophy has intrinsic defects i.e., 12° outward rotation of the posterior pelvis, 18° external rotation and 30% shortening of the pubic ramii.[7] Osteotomy combined with fixation of the pelvis allows the posterior urethra, bladder neck, bladder and abdominal wall to heal appropriately. The decreased stress on the midline closure reduces the incidence of wound dehiscence. Approximation of the pubic ramii also permits reconstitution of the pelvic floor bringing the levators and puborectalis sling together. This may have a beneficial effect on later continence.

The current approach is the use of a combined vertical and horizontal pelvic osteotomy for the both primary and reoperative exstrophy closure.[8] Anterior iliac osteotomy is performed to correct the ischiopubic deformity and a posterior vertical iliac osteotomy is performed to correct the external rotation of the iliac wings. Healing of the osteotomies is encouraged with the use of fixator pins and minimal skin traction for 6 weeks. If initial pubic diastasis is not completely corrected at the initial procedure, gradual correction by medial rotation of the fixator device and pins can be performed at the bedside with appropriate sedation.

Combined Bladder Closure and Epispadias Repair

The addition of epispadias repair in the older patient undergoing bladder closure allows for reduction in the number of operative procedures without compromising outcomes, prevents any degree of bladder prolapse, allows penile lengthening at the time of osteotomy and may help to increase bladder capacity.[9] The addition of preoperative testosterone stimulation leads to enhanced penile size and increases available skin for penile resurfacing.

Treatment of Outlet Obstruction

Following bladder closure, close monitoring of the upper tracts with ultrasonography, is essential to allow early diagnosis of outlet obstruction. Once the diagnosis of outlet obstruction is made, treatment should consist of cystoscopic evaluation and dilatation or endoscopic incision as indicated. Intermittent catheterization my be used to drain the bladder and to continue urethral dilatation. Vesicostomy and bladder defunctionalization are not recommended. Urinary diversion and continent pouches are considered a last resort in this instance.

OVERALL RESULTS

The use of modern surgical techniques and appropriate orthopedic intervention have allowed most children presenting with failed initial closure to be candidates for reconstruction. Of 40 patients referred following initial failed closure, all have stayed closed following reclosure at Johns Hopkins with minimum 24 months follow-up.[10] Complications noted in 10 patients included temporary femoral nerve palsy (6), urethral stricture (1), urethrocutaneous fistula (1), epididymitis (1), and urinary retention (1). Later continence however remains in doubt. In patients undergoing more than two closures, the ability of the bladder to gain adequate capacity for bladder neck reconstruction is severely compromised. Reclosure of the bladder following initial failure is performed with the hope that adequate capacity will develop in some patients to permit later bladder neck reconstruction. In those patients who do not acquire adequate capacity, increase in the size of the bladder plate will provide a more suitable template for later augmentation or reconstruction.

CONCLUSION

Prevention of failure of the primary closure remains the best approach. Following failure of initial bladder closure, subsequent attempts at reclosure should be delayed at least 4–6 months. Patients failing initial bladder closure have a significantly reduced capacity to attain adequate capacity for later bladder neck reconstruction. Those attaining adequate bladder capacity have a low probability of achieving spontaneous voiding. It is difficult to predict which patients will develop adequate capacity for later bladder neck reconstruction and so the current approach at Johns Hopkins is to reclose all patients who have an adequate bladder plate.

REFERENCES

1. Oesterling JE, Jeffs RD: The importance of successful initial closure in the surgical management of classical bladder exstrophy: analysis of 144 patients treated at Johns Hopkins Hospital between 1975 and 1985. J Urol., 137:258–262 1987.
2. Gearhart JP, Ben-Chaim J, Sciortino C, Sponsellor PD, Jeffs RD: The multiple reoperative bladder exstrophy closure: What affects the potential of the bladder? Urology 47(2):240–243 1996.
3. Husmann DA, McLorie GA, Churchill BM: Closure of the exstrophic bladder: evaluation of the factors leading to success and its importance in urinary continence. J Urol., 142:522–525 1989.
4. Gearhart J.P.: Bladder and urachal abnormalities: The exstrophy-epispadias complex. In: Clinical Pediatric Urology 3rd ed. Eds. Kelalis P.P., King L.R., Belman A.B.; WB Saunders Co p 579.
5. Gearhart JP, Peppas DS, Jeffs RD: The failed exstrophy closure: Strategy for management. Br J Urol., 71:217–220 1993.
6. Gearhart JP, Jeffs RD: Management of the failed exstrophy closure. J Urol., 146:610 1991.
7. Sponsellor PD, Bisson LS, Gearhart JP, Jeffs RD, Majid D, Fishman E: The anatomy of the pelvis in the exstrophy complex. J Bone Joint Surg., 77:177 1995.
8. Gearhart JP, Forschner DC, Jeffs RD, Ben-Chaim J, Sponsellor PD: A combined vertical and horizontal pelvic osteotomy approach for primary and secondary repair of bladder exstrophy. J Urol., 155:689–693 1996.
9. Mathews R, Taylor S, Jeffs RD, Gearhart JP: Combined bladder closure and epispadias repair in the management of classic bladder exstrophy. Submitted for publication J Urol. 1997.
10. Gearhart J.P. Jeffs R.D.: State-of-the-art reconstructive surgery for bladder exstrophy at the Johns Hopkins Hospital. Am J Dis Child 143:1475, 1989.

16

APPLICATIONS OF URETEROSIGMOIDOSTOMY IN BLADDER EXSTROPHY

Raimund Stein,* Margit Fisch, and Rudolf Hohenfellner

Department of Urology
Johannes Gutenberg University, Medical School
Mainz, Germany

ABSTRACT

Until 1994, 95 patients with bladder exstrophy and 20 with incontinent epispadias were operated upon at our department. During the last years more and more exstrophy-cripples were referred to our institution after primary treatment elsewhere. Thus a total of 56 of the 104 patients had a secondary treatment after failure or unsatisfactory treatment. In contrast to many other institutions, our standard procedure was primary ureterosigmoidostomy and, since 1990 it is completely replaced by a modification of it, the sigma rectum pouch (Mainz Pouch II). After failed primary bladder closure and in patients with short or pathological ureters or insufficient anal sphincter function, we prefer an ileocaecal pouch (Mainz Pouch I). In patients with a dilated upper urinary tract we perform an colonic conduit with the option of conversion into a continent urinary diversion.

104 patients could be followed, two of whom died in the meantime, of causes not related to urinary diversion. Mean followup after the first operation was 16.7 years.

At the latest follow-up 52 had a rectal reservoir, 31 a Mainz pouch I, 15 a colonic conduit and 6 a sling plasty (incontinent epispadias) or modified Young Dees procedure. Upper urinary tract remained stable in 98%, 95%, 96% and 100% respectively. Complete day and night time continence: rectal reservoirs 94%, Mainz pouch I 97% and mod. Young Dees/sling plasty 50%.

As the sigma rectum pouch (Mainz pouch II) is the primary treatment of choice at our institution, this report will focus on our clinical experience in detail. Between 1990

* Contact author: Dr. Raimund Stein, Langenbeck Str. 1, 55131 Mainz, Germany, Tel. 0049 6131 177177, Fax: 0049 6131 232986 or 230462

The Exstrophy–Epispadias Complex, edited by Gearhart and Mathews.
Kluwer Academic / Plenum Publishers, New York, 1999.

and 1994 the Mainz pouch II was performed in 87 patients (20% with bladder exstrophy or incontinent epispadias). Early complication rate was 6.9%, in 9.6% late complications were followed by intervention. The overall rate of stenosis at ureteral implantation was 7.2%. 94% of the patients were continent day and night.

INTRODUCTION

Achieving urinary continence in the management of patients with the exstrophy-epispadias complex remains a therapeutic challenge. The optimal type of urinary tract reconstruction is still a matter of debate. In most institutions in the United States and Europe, the operative procedure of choice is primary bladder closure and bladder neck reconstruction. However, the risk of damage to the upper urinary tract is associated with this procedure and the continence rate reported in the literature varies from 32% to 80%.[1,2] Multiple procedures are often required to achieve continence with a reoperation rate of up to 75%.[3,4]

In contrast to many other institutions, our method of choice for more than 25 years has been primary urinary diversion in the form of ureterosigmoidostomy, which has shown a complete continence rate of 92.3%.[5] Since 1990 the procedure has been modified to include a sigma rectum pouch—the Mainz Pouch II—offering the advantage of a low-pressure reservoir.[6]

PATIENTS AND METHOD

Between 1968 and 1994, 95 patients with bladder exstrophy and 20 with incontinent epispadias were operated upon at our institution. Follow-up was possible in 104 of the 115 patients, 9 had moved out of the country, one was lost to followup and one patient with serious family problems refused participation. Two further patients died of causes not related to urinary diversion.

Of the 104 patients followed in 48 patients primary treatment was performed, a rectal reservoir in 38, including 3 patients who received only a genital reconstruction at our institution. A total of 56 patients were referred to our institution after primary treatment elsewhere.

Out of 87 patients with a sigma rectum pouch (operated between 1990 and 1994) 20% had bladder exstrophy or incontinent epispadias.

The followup data concerns renal function, continence, social development and late complications. Continence was defined as use of the toilet at normal intervals, voluntary control of the urine stream, wearing regular underclothes and no need for pads.

RESULTS

Of the 56 patients with unsuccessful or unsatisfactory primary treatment performed elsewhere in 30 (53%) had a Mainz pouch I at the latest follow-up. Serum creatinine levels of all patients were within normal range, the upper urinary tract remained stable in 95% of all renal units. 29 of the 30 patients were completely continent (97%). In order to preserve renal function after failure of urinary tract reconstruction a colon conduit was performed in 12 patients (21%). One boy has a slight dilatation of the right upper tract, in the remaining patients the upper tract remained stable or improved during the followup period.

Up to 1994 a total of 87 patients received a Mainz pouch II, 17 of them had a bladder exstrophy or incontinent epispadias (20%). 7 early complications were encountered in 6 patients (6.7%), only two of them were pouch related. Late complications followed by intervention occurred in 8 patients (8.9%). The overall rate of stenosis at the ureteral implantation was 7.2%. 94% of the patients were continent at night. The majority of the patients take alkalinizing drugs to avoid metabolic acidosis.

DISCUSSION

In patients with bladder exstrophy or incontinent epispadias, complete urine continence and preservation of the upper urinary tract has to be the primary aim of treatment, as continence is essential for these children to achieve good social integration.

Primary bladder closure with or without osteotomy in the first 72 hours of life followed by bladder neck reconstruction (Young Dees) at the age of 2 1/2 to 5 is still the most frequently used form of continence surgery for patients with the exstrophy-epispadias complex.[7] The reported continence rate is up to 80%,[2] and with a well-selected patient collective in the hands of experienced surgeons, it could be even higher. However this is in contrast to the fact, that during the last years more and more exstrophy-cripples were referred to our institution after primary bladder closure performed elsewhere.

In addition, in the literature, the term continence appears to be defined differently.[4,7–10] Mesrobian et al. define "continent" as the ability to remain dry for three or more hours. Occasional episodes of diurnal or nocturnal incontinence are not considered to represent failure in achieving continence.[10] Jones et al. define continence as totally dry day and night for more than four hours, wearing regular underclothes without protective padding. However, with this definition, complete continence following the standard Young-Dees-Leadbetter bladder neck reconstruction was only achieved in 39%, with a modified procedure in 64% of their cases.[4] At the meeting of the American Academy of Paediatric Urology in 1992, Jeffs reported that in their group of 5 to 10-year-old patients, only 20% were completely dry. "Most children have a 3-hour dry interval and we try to get them to use that period to plan getting from bathroom to bathroom". During the night they have to wear diapers or have to get up one or more times to use the toilet.[9] Furthermore, Hollowell's and Ransley's collective of continent patients—for example—includes children who can only empty their bladder by clean intermittent catheterisation (CIC).[8] Especially in boys, CIC involves a high risk of epididymitis, urethral strictures and even perforation of the augmented bladders.[11, 12]

In our opinion the definition of continence in patients with the exstrophy-epispadias complex should not be different from that of the normal population, using the toilet in normal intervals, voluntary control of the urine stream, wearing regular underclothes and no need for pads.

One reason for failure of bladder closure and bladder neck reconstruction to achieve complete continence is the fact that the creation of a functional sphincter mechanism is ultimately impossible. To achieve continence, additional procedures beyond a Young-Dees-Leadbetter bladder neck reconstruction often become necessary and include intestinal bladder augmentation, pubovaginal sling, artificial urinary sphincter replacement and ureter reimplantation.[3,4,10,13–15] All of these children spend long periods of time in the hospital, away from their home, friends and school.

A further reason may be insufficiency of the detrusor muscle itself. Urodynamic studies by Hollowell et al. demonstrate urine leakage due to involuntary detrusor contractions in 10 out of 18 patients with an unsatisfactory state of continence. In their continent group 5 of the 7 patients could not void effectively and required intermittent catheterisation to empty completely. Only 8% demonstrated normal bladder function during filling and contractile voiding. The authors conclude that "evolution to completely normal lower urinary tract function after bladder neck reconstruction is an unrealistic expectation for the majority of children with exstrophy and severe epispadias."[8] Further research is needed to determine whether the detrusor muscle has or could achieve normal function and if treatment with anticholinergic drugs would be helpful. Early results of our own research showed that spontaneous contractions of the detrusor muscle were not influenced by anticholinergic drugs.[16]

Preservation of the upper urinary tract has to be kept in mind as an important aim. Mesrobian et al. showed that hydronephrosis (25%) and reflux (43%) is not uncommon in patients with primary bladder closure.[10] Furthermore, deterioration of the upper urinary tract occurred in up to 24%.[4]

The other option in treating patients with bladder exstrophy or incontinent epispadias is permanent urinary diversion. Ureterosigmoidostomy was the first form of continent urinary diversion.[17] After initial enthusiasm in the first half of the century, a high postoperative complication rate occurred and deterioration of the upper urinary tract became evident. Electrolyte imbalance, urinary tract infections, pyelonephritis, renal insufficiency and uraemia were not uncommon in the past.[10] An increasing number of reports of colonic tumors associated with ureterosigmoidostomy also reduced the frequency of this operation. Refinements in surgical technique, antireflux ureteral implantation as well as improved antibiotic treatment and suture materials have again made this standard surgical procedure an attractive form of continent urinary diversion. However, as we learned, a previously dilated upper urinary tract represents a contraindication for this procedure. Our patients as well as published reports from other institutions document that individuals who have had ureterosigmoidostomy since early childhood are satisfied with their life and do not desire any other type of urinary diversion.[18] Both demonstrate that bladder exstrophy patients with urinary diversion achieve a normal education and lead a successful social life; the unemployment rate in these adults is very low.[19–21] Incontinence is mostly associated with late urinary diversion.[18,22] In our group of children who underwent primary ureterosigmoidostomy performed the age of 6, 92% were completely continent when they entered school.

The Mainz Pouch II, a modification of ureterosigmoidostomy, offers several improvements. Antimesenteric opening of the bowel creates a low-pressure reservoir, thus preventing reflux and subsequent pyelonephritis.[6] This is advantageous for the upper urinary tract and essential for good continence results especially during the night.

After initial or secondary failure of bladder exstrophy closure, Gearhart and Jeffs recommend reclosure and additional osteotomy, even if the latter has already been performed previously. The child is immobilised by external fixation for about 8 weeks. The results the above authors achieved in 24 patients are not convincing: 7 patients were dry for > 4 hours during the day, 1 remained completely incontinent, 2 patients were dry for more than 4 hours on CIC, 1 patient with a continent stoma and bladder augmentation remained dry on CIC at 4 hour intervals, one patient required a colonic conduit and 12 patients were awaiting further operations.[13] In contrast to this strategy we prefer conversion to a continent urinary diversion after failed bladder closure and/or bladder neck reconstruction. In patients with sufficient anal sphincter function and a normal upper urinary tract we prefer the Mainz Pouch II. The ureteral implantation technique of Abol Enein using the serous lined extramural tunnel can be transferred to the sigma rectum pouch, thus moderate dilated ureters can be implanted into rectal reservoirs.[23,24] However, due to the previously failed operations the ureters are severely dilated and/or too short to perform a Mainz Pouch II. In patients with insufficient anal sphincter function the Mainz Pouch I procedure is carried out. This type of urinary diversion includes a low-pressure reservoir with a large capacity.[25] In this group of patients with a Mainz Pouch I high continence rates (97%) are achieved with one operation.[26]

All patients with intestinal urinary diversion run the risk of metabolic disturbances probably resulting in late complications.[27] If colon tissue is used to create the urinary diversion, electrolyte disturbances may occur. Careful monitoring of the acid-base balance is necessary to avoid complications. Early correction with alkalinising agents (sodium bicarbonate and/or lemon juice) are recommended if the base excess is below -2.5. With this strategy, none of our patients developed a metabolic acidosis.

The other important late complication is the development of secondary malignancies. Most of these reports of neoplasms in urinary diversion refer to ureterosigmoidostomy.[28–31] However, concerning the number of reports, one thing should be kept in mind: this is the oldest type of urinary diversion with the longest period of followup. The risk of colon carcinoma after ureterosigmoidostomy has been calculated to be about 5%.[29,30] This is about 150–500 times higher than the normal population, but does not evaluate the operative technique nor the geographically and racially correlated incidence of colon carcinoma.[28,30,31] Hence, screening of all patients with urinary diversion is mandatory. A colonoscopy has to be performed 5 years after ureterosigmoidostomy or Mainz Pouch II and should be repeated annually thereafter. The mean followup of all our patients, including those in whom conversion to ureterosigmoidostomy was performed, is now 18.8 years. To date, we have not seen any malignancies in individuals with bladder exstrophy after ureterosigmoidostomy. In our total group of patients with ureterosigmoidostomy (n=198) we have seen 3 tumors at the implantation site.

Malignancy has also been reported in the retained exstrophied bladder.[10] It remains unclear, if the risk of carcinoma is as high as in patients with urinary diversion using bowel tissue. In the patients with closure of the exstrophied bladder the followup period has been too short compared to that of ureterosigmoidostomy patients. There is reason to assume that the incidence of carcinoma in bladder augmentation using bowel tissue may not be decisively different from that in ureterosigmoidostomy. All forms of urinary tract reconstruction using bowel segments are associated with an increased risk of secondary malignancies.[28]

CONCLUSION

Patients with continent forms of urinary diversion are generally continent day and night. This is the standard which reconstructive surgery of primary bladder closure should achieve. In view of the continence rates achieved by primary bladder closure and the urodynamic studies in these patients this seems to be more than doubtful. More and more children and young adults with incontinence and deterioration of the upper urinary tract were seen at our institution after primary bladder closure performed elsewhere with the wish for urinary diversion. Thus, the concept of bladder exstrophy management should be reconsidered as well as redefinition of the term "continence" for such operations is not valid. Furthermore, none of our patients developed renal insufficiency or has presently a deterioration of the upper urinary tract. Continent urinary diversion performed in early childhood saves these patients repeated operations as well as physical and psychological sequelae. The first operation in these patients must be the best one, as it determines their fate.

REFERENCES

1. J. S. Ansell: Surgical treatment of exstrophy in bladder with emphasis on neonatal primary closure: Personal experience with 28 consecutive cases treated at the university of Washington Hospital from 1962 to 1977: Techniques and results. Journal of Urology, 650653, 1979.
2. D. A. Husmann, G. A. McLorie and C. B. M.: Closure of the exstrophic bladder. Journal of Urology, **142**: 522–527, 1989.
3. J. P. Connor, T. W. Hensle, J. K. Lattimer and K. A. Burbige: Long-term followup of 207 patients with bladder exstrophy: An evolution in treatment. Journal of Urology, **142**: 793–796, 1989.
4. J. A. Jones, M. E. Mitchell and R. C. Rink: Improved results using a modification of the Young-Dees-Leadbetter bladder neck repair. British Journal of Urology, **71**: 555–561, 1993.
5. M. Stöckle, E. Becht, G. Voges, H. Riedmiller and R. Hohenfellner: Ureterosigmoidostomy: An outdated approach to bladder exstrophy? Journal of Urology, **143**: 770–775, 1990.
6. M. Fisch, R. Wammack, S. C. Müller and R. Hohenfellner: Mainz Pouch II (Sigma rectum pouch). Journal of Urology, **149**: 258–264, 1993.
7. J. P. Gearhart: Editorial comments. Journal of Urology, **143**: 774, 1990.
8. J. G. Hollowell, P. D. Hill, P. G. Duffy and P. G. Ransley: Evaluation and treatment of incontinence after bladder neck reconstruction in exstrophy and epispadias. British Journal of Urology, **71**: 743–749, 1993.
9. R. D. Jeffs: Discussion at the AAP, 1992, San Francisco. J. Urol., **150**: 634, 1993.
10. H. G. Mesrobian, P. P. Kelalis and S. A. Kramer: Long-term followup of 103 patients with bladder exstrophy. Journal of Urology, **139**: 719–722, 1988.
11. H. G. Rushton, J. R. Woodard, T. S. Parrott, R. D. Jeffs and J. P. Gearhart: Delayed bladder rupture after augmentation enteroplasty. Journal of Urology, **140**: 344–346, 1988.
12. J.-J. Wyndaele and D. Maes: Clean intermittent self-catheterisation: A 12-year followup. Journal of Urology, **143**: 906–910, 1990.
13. J. P. Gearhart and R. D. Jeffs: Management of the failed exstrophy closure. Journal of Urology, **146**: 610–612, 1991.
14. J. P. Gearhart, C. J. Ben, C. Sciortino, P. D. Sponseller and R. D. Jeffs: The multiple reoperative bladder exstrophy closure: what affects the potential of the bladder? Urology, **47**: 240–3, 1996.
15. J. G. Hollowell and P. G. Ransley: Surgical management of incontinence in bladder exstrophy. Br. J. Urol., **68**: 543 - 548, 1991.
16. R. Wammack, M. Fisch, S. C. Müller, H. Nawrath and R. Hohenfellner: Pharmacology, electrophysiology and immunhistochemestry of the exstrophic bladder. ESPU Göteborg, Abstract 44, 1994.
17. J. Simon: Ectopia vesica (Absence of the anterior walls of the bladder and pubic abdominal parities); operation for directing the orifices of the ureter into the rectum; temporary success; subsequent death; autopsy. Lancet, **2**: 568–570, 1852.
18. S. H. Silverman, C. R. J. Woodhouse, J. R. Strachan, J. Cumming and M. R. B. Keighley: Long-term management of patients who have had urinary diversion into colon. British Journal of Urology, **58**: 634, 1986.
19. R. Stein, M. Fisch, M. Stöckle and R. Hohenfellner: Urinary Diversion in Bladder Exstrophy and Incontinent Epispadias: 25 years of experience. J. Urol., **154**: 1177 - 1181, 1995.
20. C. R. J. Woodhouse: Exstrophy and epispadias. In: Long-term paediatric urology. Edited by C. R. J. Woodhouse. Oxford, England: Blackwell Scientific Publications, pp. 127 - 150, 1991.
21. A. Zabbo and R. Kay: Ureterosigmoidostomy and bladder exstrophy: A long-term follow-up. Journal of Urology, **136**: 396–398, 1986.
22. M. T. Macfarlane, J. K. Lattimer and T. W. Hensle: The unheralded hazard of ureterosigmoidostomy. Paediatrics, **64**: 668 - 671, 1976.
23. E. H. Abol and M. A. Ghoneim: A novel uretero-ileal reimplantation technique: the serous lined extramural tunnel. A preliminary report. J. Urol., **151**: 1193 - 1197, 1994.
24. M. Fisch, H. Abol Enein and R. Hohenfellner: Ureterimplantation mittels serösem extramuralem Tunnel in Mainz-Pouch I und Sigma-Rektum-pouch (Mainz-Pouch II). Akt. Urol., **26**: I-X, 1995.

25. J. W. Thüroff, P. Alken, H. Riedmiller, G. H. Jacobi and R. Hohenfellner: 100 cases of Mainz Pouch: continuing experience and evolution. Journal of Urology, **140**: 283–288, 1988.
26. R. Stein, M. Fisch, M. Stockle and R. Hohenfellner: Urinary diversion in bladder exstrophy and incontinent epispadias: 25 years of experience. J Urol, **154**: 1177–81, 1995.
27. W. S. McDougal: Metabolic complications of urinary intestinal diversion. Journal of Urology, **147**: 1199–1208, 1992.
28. T. Kälble, S. W., M. R. Berger, R. Waldherr, F. Amelung, K. Möhring and G. Staehler: Karzinomrisiko in verschiedenen Formen der Harnableitung unter Verwendung von Darm. Aktuelle Urologie, **24**: 1–7, 1993.
29. E. H. Pierce, P. Zickermann Jr. and G. W. Leadbetter: Ureterosigmoidostomy and carcinoma of the colon. Trans. Amer. Ass. Gentito-Urin. Surg., **70**: 92–94, 1978.
30. G. S. Sooriyaarachchi, R. O. Johnson and P. P. Carbone: Neoplasm of the large bowel following ureterosigmoidostomy. Arch. Surg., 1174–1179, 1977.
31. H. M. Spence, W. W. Hoffmann and G. P. Fosmire: Tumors of the colon as a late com-plication of ureterosigmoidostomy for exstrophy of the bladder. British Journal of Urology, **51**: 446–470, 1979.

17

CYSTOPLASTY IN BLADDER EXSTROPHY

Indications and Techniques

Steven G. Docimo

The Johns Hopkins Hospital
The James Buchanan Brady Urological Institute
Baltimore, Maryland

Despite our institutional preference for functional reconstruction of the child with bladder exstrophy,[1] the need for bladder augmentation or continent urinary diversion is not infrequent. Techniques for augmentation have not changed tremendously in the past twenty years, although there are some variations and perhaps some completely new techniques on the horizon.[2] The decision to augment the bladder in a patient with exstrophy may allow for urinary continence, but it will also cause significant changes in the patient's lifestyle. These changes are due to techniques of emptying the augmented bladder, potential complications of the augmented bladder, cosmetic implications of augmentation surgery and potential long-term metabolic and neoplastic consequences of bladder augmentation. Because these operations are generally performed in young children, these considerations must weigh heavily on reconstructive decision making.

In this manuscript I would like to review the indications for augmentation in classic bladder exstrophy, the strategies for managing the bladder plate and the bladder neck, strategies for catheterization of the augmented bladder, techniques of bladder augmentation focusing on tissues used and the surgical approaches to augmentation.

The indications for bladder augmentation in the child with classic exstrophy are different at different institutions. It is our strong bias that functional reconstruction should be attempted in nearly every case of classic bladder exstrophy. Therefore the indications for augmentation are somewhat selective. These would include failure of bladder growth, despite maneuvers such as collagen injection at the bladder neck[3] or early epispadias repair.[4] Failure of bladder neck reconstruction, especially after multiple attempts or in the bladder which is too small to seriously consider another attempt at reconstruction would be an indication for bladder augmentation in many cases.[5] Other less common indications for bladder augmentation would include manifestations of a low volume high pressure bladder such as the failure of ureteral reimplants and or recurrent urinary tract infection and a history of spontaneous bladder rupture.

The Exstrophy–Epispadias Complex, edited by Gearhart and Mathews.
Kluwer Academic / Plenum Publishers, New York, 1999.

It has not been unusual in the historical treatment of classic bladder exstrophy to perform cystectomy and manage the urinary tract with some form of supravesical diversion or ureterosigmoidostomy. This is rarely a primary mode of management in most institutions in this country because there can be several advantages to maintaining the bladder as part of the eventual urinary tract reconstruction. First, reimplanting the ureters into the bladder plate avoids ureteroenterostomies, which have a higher complication rate than ureteroneocystostomies. The bladder likewise forms a nice template for implantation of a Mitrofanoff stoma.[6] If the ureters and the continent stoma are reimplanted into the bladder, there is more flexibility in the tissues which can be used for the augmentation because implantation does not need to be performed in those tissues. Anatomically, the bladder plate serves to anchor the augment in the pelvis which can increase the ease of eventual percutaneous procedures for stone removal or allow the maintenance of the urethra in select cases.

At our institution, it is not possible to maintain the integrity of the urethra in the bladder neck in the majority of patients who come to bladder augmentation.[7] First of all, many of these patients have failed bladder neck reconstruction. The entire bladder can be used as a bladder neck as has been described by Arap,[8] but this can be difficult to catheterize especially in line with a reconstructed epispadias and it obviates the advantage of having a template for ureteral reimplantation. Most often the bladder neck is divided to provide for continence in this area. Division of the bladder neck has some disadvantages including the lack of a "pop off valve" which might help reduce the incidence of spontaneous augmentation perforation.[9] In addition, one has to consider where the seminal fluid will drain, especially if there is urethral stricture or non continuity. Finally, one might consider fascial slings or artificial sphincters, techniques which have been used successfully in neurogenic incontinence and other forms of urethral insufficiency. Unfortunately, after bladder neck reconstruction there is poor success and high erosion risk for these procedures.[10] Therefore they are rarely useful in the patient with bladder exstrophy.

The tissues which are currently used for bladder augmentation can be divided into two broad groups. These are urothelial line tissue and gastrointestinal tissue. Urothelial line tissue would include bladder auto augmentation,[11] auto augmentation with the use of demucosalized stomach or intestine over the bladder mucosa[12–14,15] or augmentation of the bladder with native ureter.[16,17] Gastrointestinal tissues commonly used for bladder augmentation includes stomach, ileum and colon.

There are clear and inherent advantages to using urothelial tissues for bladder augmentation[2]. First, these are tissues which are developed to be in contact with urine. There is therefore no difficulty with absorption of electrolytes from the urine nor is there excess mucous production. Unfortunately, auto augmentation techniques with or without gastrointestinal seromuscular backing are rarely applicable to the exstrophy bladder because of the small size of the bladder and its mucosa. Ureteral augmentation can be advantageous and has been frequently used although its application is limited by the availability of dilated ureter, the need to drain an intact kidney by transuretero-ureterostomy which may not leave enough distal ureter for sufficient augmentation, and the unknown expansion properties of a ureteral augment at the time of surgery. Despite these disadvantages, ureteral augmentation should be considered where possible.

The use of stomach, ileum and colon in the urinary tract has been well documented over many years.[18] Although we have come to take the disadvantages of the use of these tissues somewhat for granted, in light of the potential of future techniques it is worth reviewing some of these draw backs. Both ileum and colon are absorptive tissues and create a tendency toward hyperchloremic metabolic acidosis. Most patients with bladder exstrophy have normally functioning kidneys and this is not a significant concern. With renal impairment, the likelihood of developing significant acidosis becomes greater. Both ileum and colon produce significant amounts of mucous requiring larger catheters for bladder emptying and daily irrigation of the pouch. The mucous is probably also a significant contributor to stone formation which occurs anywhere from 15 to 50% of bladders augmented with ileum and colon.[19,20] Growth retardation due to the use of ileum and colon in the bladder has been suggested in previous studies although this has not been corroborated.[21] The risk of spontaneous bladder perforation exists with all forms of bladder augmentation, but was first described in gastrointestinal augments.[22–24] Finally, removing significant amounts of terminal ileum or colon from the gastrointestinal tract can result in diarrhea, malabsorption and potentially a tendency toward kidney stone formation.[25–27]

Stomach as a substrate for bladder augmentation has some advantages over intestine. First of all, in patients with renal impairment the stomach actually assists in excretion of acids and thereby ameliorates the underlying metabolic acidosis which would be exacerbated by colon or ileum.[28] Second, the stomach is amenable to tunneling much like a bladder, making it ideal for ureteral or Mitrofanoff implantation. There are, however, risks of gastrocystoplasty which have been established over the past decade. The most common complication of stomach in the bladder is the hematuria-dysuria syndrome.[29] Up to half of gastrocystoplasty patients will experience hematuria or bladder pain, but most can be managed expectantly, and the majority of others can be treated with H2 blockers or omeprazole with resolution.[30] Bladder perforation due to ulceration occurs in the face of oliguria, and makes gastrocystoplasty a questionable choice for the patient in end-stage renal failure. Finally, metabolic alkalosis in the face of dehydration and vomiting can at times be life threatening, and has resulted in the need to resect the bladder augmentation.[30]

One consideration when performing bladder reconstruction in young children is the likelihood that they will live with exogenous tissue in the urinary tract for many years. It is unknown at this time what the risk of neoplasia will be in augmented bladders. There have been several reported adenocarcinomas which usually occur at the interface between the intestine and bladder.[31] Why these tumors occur and how significant this risk might be is as yet unknown. Additionally, there is no current recommendation for screening of patients with augmented bladders. Clearly as the patients get to 20 or 30 years status post augmentation, this neoplastic tendency will declare itself, and some form of screening will likely need to be adopted.

One has to assume that any patient whose bladder is augmented will require intermittent catheterization for emptying. Although this is not always the case, if patients are instructed to expect catheterization up front, there is no potential for disappointment after reconstruction. Catheterizing the urethra in patients with bladder exstrophy can be difficult in two respects. It is often difficult to negotiate the reconstructed epispadiac urethra, and these urethras have full sensation so there is discomfort associated with catheterization.[1] Secondly, it is potentially difficult to maintain continence per urethra in the face of intermittent catheterization through a reconstructed bladder neck. For these reasons, most of these patients will require a catheterizable stoma. These have been constructed using many techniques over the years, but most groups would agree that the Mitrofanoff principle is currently the most reliable technique.[6,32] Mitrofanoff stomas are usually constructed of appendix if available, although the ureter, ureteral free graft, fallopian tube and stomach have all been used in this capacity.

In order to create a Mitrofanoff stoma, there needs to be a tunneled implantation to create the flap valve mechanism. The bladder base forms an excellent substrate for such an implantation, but is not always available. The Mitrofanoff stoma can be implanted into the taenia or submucosally elsewhere in the colon. As mentioned previously, submucosal tunneling is relatively easy to do in the stomach as well. The ileum is less reliable as a base for implantation. Occasionally a colon tube can be reimplanted into colon or a stomach tube can be reimplanted into a segment of stomach in order to achieve both the stoma and augmentation from one segment of the gastrointestinal tract.[33]

When creating a catheterizeable stoma, the body image of these patients must be born in mind. Although we have considered it cosmetic in the past to have a small pink stoma which can be covered by a Band-Aid, there is a clear preference among patients for a stoma which is hidden in the umbilicus if possible. This could be made easier in the exstrophy patient with a tubed umbilicoplasty,[34] or using a posterior umbilical flap as has been previously described.[35] Either of these techniques will result in a stoma which is generally completely hidden and does not require any covering.

Bladder augmentation in patients with classical exstrophy has been reviewed in the Johns Hopkins Series. The experience can be summarized as follows: There are 61 patients reviewed, 52 with classic bladder exstrophy and 9 with cloacal exstrophy. 30 patients underwent ileocystoplasty, 11 had colonic segments used, and 20 had reconstruction using ileocecal segments. 43 catheterize through continent stomas, and 18 catheterize through the urethra. Two of these 18 are incontinent. All of the 43 patients who catheterize via stoma have had the bladder neck closed. Of 63 refluxing ureters, 61 have been reimplanted, either into the bladder or the augment, and 2 have persistent reflux. The dry intervals are from 2.5 to 4 hours, with volumes from 200 to 600 ml.

The final issue to be considered is the surgical approach used for bladder augmentation or reconstruction. Classically, a midline abdominal incision has been used to allow work deep in the pelvis as well as the manipulation of gastrointestinal tissues as high as the epigastrium. Occasionally an isolated ileal or sigmoid augmentation can be done through a Pfannenstiel incision allowing for improved cosmesis and postoperative pain management. As previously reported, laparoscopic bladder augmentation using stomach or other tissues is possible,[36] although this remains technically difficult and is not available at most centers. An approach is being tested at our institution which uses laparoscopy to "harvest" the appendix and the gastrointestinal tissues to be used for reconstruction, followed by a Pfannenstiel incision through which reimplantation, bladder neck procedures and "assembly" of harvested tissues can be carried out. This results in a significantly improved cosmetic appearance for these patients who are very conscious of body image, and may result in more rapid postoperative recovery and decreased requirement for pain medication.

In summary, bladder augmentation is often required in order to achieve continence and independence in the patient with bladder exstrophy. It has been the feeling at this institution that bladder augmentation is often preferable to supravesical diversion whether continent or incontinent. Division of the bladder neck is often necessary to achieve the desired result. The continent catheterizable stoma is most frequently developed using the Mitrofanoff principle. Bladder augmentation should be carried out with urothelium if possible, but stomach, colon and ileum are all proven tissues for bladder augmentation. All of these tissues carry specific risks including potentially the risk for neoplasia in the future. Finally, some consideration should be made for minimally invasive incisions and techniques which might improve postoperative recovery and clearly confer a cosmetic advantage in this challenging group of patients.

REFERENCES

1. Docimo, S.G., Gearhart, J.P., and Jeffs, R.D.: Bladder Exstrophy, in The Bladder, Edited by J.M. Fitzpatrick and Krane, R.J. Edinburgh: Churchill Livingstone: p. 501, 1995.

2. Docimo, S.G.: Options for bladder augmentation wihtout the use of bowel mucosa. Dial Ped Urol, 18: 4, 1995.
3. Caione, P., Lais, A., de, G.M., and Capozza, N.: Glutaraldehyde cross-linked bovine collagen in exstrophy/epispadias complex. J Urol, 150: 631, 1993.
4. Gearhart, J.P. and Jeffs, R.D.: Bladder exstrophy: increase in capacity following epispadias repair. J Urol, 142: 525, 1989.
5. Gearhart, J.P., Canning, D.A., and Jeffs, R.D.: Failed bladder neck reconstruction: options for management. J Urol, 146(4): 1082, 1991.
6. Mitrofanoff, P.: Neurogenic bladder in children. Current status of our knowledge and treatment by transappendiceal continent cystostomy. [French]. Bull Soc Sci Med Grand Duche Luxemb, (219), 1987.
7. Gearhart, J.P., Peppas, D.S., and Jeffs, R.D.: The application of continent urinary stomas to bladder augmentation or replacement in the failed exstrophy reconstruction. Br J Urol, 75(1): 87, 1995.
8. Arap, S., Martins, G.A., Menezes, de, Goes, and G: Initial results of the complete reconstruction of bladder exstrophy. Urol Clin North Am, 7(2): 477, 1980.
9. Bauer, S.B., Hendren, W.H., Kozakewich, H., Maloney, S., Colodny, A.H., Mandell, J., and Retik, A.B.: Perforation of the augmented bladder. J Urol, 148: 699, 1992.
10. Gearhart, J.P. and Jeffs, R.D.: Augmentation cystoplasty in the failed exstrophy reconstruction. J Urol, 139(4): 790, 1988.
11. Cartwright, P.C. and Snow, B.W.: Bladder autoaugmentation: early clinical experience. J Urol, 142: 505, 1989.
12. Dewan, P.A. and Byard, R.W.: Autoaugmentation gastrocystoplasty in a sheep model. Br J Urol, 72(1): 56, 1993.
13. Dewan, P.A. and Stefanek, W.: Autoaugmentation gastrocystoplasty: early clinical results. Br J Urol, 74(4): 460, 1994.
14. Nguyen, D.H., Mitchell, M.E., Horowitz, M., Bagli, D.J., and Carr, M.C.: Demucosalized augmentation gastrocystoplasty with bladder autoaugmentation in pediatric patients. J Urol, 156(1): 206, 1996.
15. Gonzalez, R., Buson, H., Reid, C., and Reinberg, Y.: Seromuscular colocystoplasty lined with urothelium: experience with 16 patients. Urology, 45(1): 124, 1995.
16. Bellinger, M.F.: Ureterocystoplasty: a unique method for vesical augmentation in children. J Urol, 149(4): 811, 1993.
17. Dewan, P.A., Nicholls, E.A., and Goh, D.W.: Ureterocystoplasty: an extraperitoneal, urothelial bladder augmentation technique. Eur Urol, 26(1): 85, 1994.
18. Mitchell, M. and Burns, M.W.: Urinary undiversion and augmentation cystoplasty., in Clinical Pediatric Urology, Edited by P.P. Kelalis, King, L.R., andBelman, A.B. Philadelphia: W.B. Saunders Company: p. 904, 1992.
19. Blyth, B., Ewalt, D.H., Duckett, J.W., and Snyder, H.d.: Lithogenic properties of enterocystoplasty. J Urol, 148: 575, 1992.
20. Palmer, L.S., Franco, I., Kogan, S.J., Reda, E., Gill, B., and Levitt, S.B.: Urolithiasis in children following augmentation cystoplasty. J Urol, 150: 726, 1993.
21. Mundy, A.R. and Nurse, D.E.: Calcium balance, growth and skeletal mineralisation in patients with cystoplasties. Br J Urol, 69(3): 257, 1992.
22. Sheiner, J.R. and Kaplan, G.W.: Spontaneous bladder rupture following enterocystoplasty. J Urol, 140: 1157, 1988.
23. Rushton, H.G., Woodard, J.R., Parrott, T.S., Jeffs, R.D., and Gearhart, J.P.: Delayed bladder rupture after augmentation enterocystoplasty. J Urol, 140(2): 344, 1988.
24. Elder, J.S., Snyder, H.M., Hulbert, W.C., and Duckett, J.W.: Perforation of the augmented bladder in patients undergoing clean intermittent catheterization. J Urol, 140: 1159, 1988.
25. Davidsson, T., Lindergard, B., and Mansson, W.: Long-term metabolic and nutritional effects of urinary diversion. Urology, 46(6): 804, 1995.
26. Steiner, M.S., Morton, R.A., and Marshall, F.F.: Vitamin B12 deficiency in patients with ileocolic neobladders. J Urol, 149(2): 255, 1993.
27. Canning, D.A., Perman, J.A., Jeffs, R.D., and Gearhart, J.P.: Nutritional consequences of bowel segments in the lower urinary tract. J Urol, 142: 509, 1989.
28. Adams, M.C., Mitchell, M.E., and Rink, R.C.: Gastrocystoplasty: an alternative solution to the problem of urological reconstruction in the severely compromised patient. J Urol, 140: 1152, 1988.
29. Nguyen, D.H., Bain, M.A., Salmonson, K.L., Ganesan, G.S., Burns, M.W., and Mitchell, M.E.: The syndrome of dysuria and hematuria in pediatric urinary reconstruction with stomach. J Urol, 150: 707, 1993.
30. Gosalbez, R.J., Woodard, J.R., Broecker, B.H., and Warshaw, B.: Metabolic complications of the use of stomach for urinary reconstruction. J Urol, 150: 710, 1993.

31. Gregoire, M., Kantoff, P., and DeWolf, W.C.: Synchronous adenocarcinoma and transitional cell carcinoma of the bladder associated with augmentation: case report and review of the literature. [Review]. J Urol, 149(1): 115, 1993.
32. Elder, J.S.: Continent appendicocolostomy: a variation of the Mitrofanoff principle in pediatric urinary tract reconstruction. J Urol. 148(1): 117, 1992.
33. Gosalbez, R., Padron, O.F., Singla, A.K., Woodard, J.R., and Galloway, N.T.: The gastric augment single pedicle tube catheterizable stoma: a useful adjunct to reconstruction of the urinary tract. J Urol, 152: 2005, 1994.
34. Sumfest, J.M. and Mitchell, M.E.: Reconstruction of the umbilicus in exstrophy. J Urol, 151(2): 453, 1994.
35. Ben-Chaim, J., Rodriguez. R., and Docimo, S.G.: Concealed umbilical stoma: description of a modified technique. Journal of Urology, 154(3): 1169, 1995.
36. Docimo, S.G., Moore, R.G., Adams, J., and Kavoussi, L.R.: Laparoscopic bladder augmentation using stomach. Urology, 46(4): 565, 1995.

18

BLADDER EXSTROPHY

Therapeutic Alternatives

Sami Arap and Amilcar M. Giron

Divisão Clínica Urológica
Hospital das Clínicas da Faculdade de Medicina USP
Caixa Postal 11273-9
05422-970 São Paulo-SP, Brazil

INTRODUCTION

Bladder exstrophy is the most common defect among the exstrophic anomalies, which presents a wide spectrum varying from cloacal exstrophy to distal penile epispadias. The child is affected in several ways, involving the genitourinary tract, musculoskelectal system and also the terminal portion of the intestinal tract. Pubic diastasis is the stigma of a spectrum of bladder exstrophy-epispadias complex malformations.

Treatment of bladder exstrophy should provide socially adequate urinary continence, protection of renal function and genital reconstruction with preservation of sexual function. In a review of the literature more than 50 different procedures were mentioned for treatment and there is no ideal nor even consistently satisfactory treatment currently available.

This assertion is perhaps less real today than it was in the 60's, when it was pronounced.

In our institution several different procedures have been utilized (Table 1).

Historical trials: from 1960 to 1968, one stage primary reconstruction, neorectal bladder, and ileal conduits were utilized with poor results (1,2,3,4,5,6).

COLOCYSTOPLASTY (ARAP PROCEDURE)

Between 1968 to 1982 we performed a procedure described as "Complete reconstruction of bladder exstrophy" (7), based on successful experience of urethro-trigonal tubularization (Young-Dees-Leadbetter technique)(8,9,10) for continence mechanism; this program was similar to that proposed by Couvelaire (11) in 1961 and Martinez Piñero (12) in 1968.

The Exstrophy–Epispadias Complex, edited by Gearhart and Mathews.
Kluwer Academic / Plenum Publishers, New York, 1999.

Table 1. Bladder exstrophy—surgical treatment

Procedure	Cases
Primary functional reconstruction	27
Ileal conduit	23
Neorectal bladder	06
Ureterosigmoidostomy	11
Colocystoplasty (Arap procedure)	56
Staged reconstruction	81

Division of Urology 1968–1996—São Paulo University Medical School

The technique was based in the following principles:

a. bladder enlargement utilizing a colonic intestinal segment;
b. antirefluxing ureterocolic anastomosis;
c. dry reconstruction of bladder and urethra;
d. plastic reconstruction of the lower abdominal wall (with or without osteotomy);
e. bladder tubularization for continence mechanism.

This strategy was devised to avoid the complications observed in bladder exstrophy one stage reconstruction, previously mentioned.

OBJECTIVES AND METHODS

Bladder exstrophy usually presents the following problems: small bladder capacity, vesicoureteral reflux with progressive deterioration of renal function (after reconstruction), urinary incontinence and surgical complications such as urinary fistulas, breakdown of sutures and re-eversion of the bladder.

The colocystoplasty or Arap Procedure (7) tried to avoid the problems described above. The following steps constitute the objectives of the method:

a. Creation of a sigmoid conduit with antireflux implantation of the ureters, which will become the "future" bladder and simultaneously is a temporary urinary diversion, allowing dry reconstruction of the bladder and urethra.
b. Iliac osteotomy in the initial cases and lately, we utilized bilateral groin flaps including skin, subcutaneous tissue and aponeurosis, to close the abdominal wall anterior defect.
c. Urethro-trigonal tubularization, as the third step, seven to ten days after osteotomy. The urethra was reconstructed according to the Young-Dees technique (8,9). The urethro-trigonal muscular tube was performed in two ways: suturing the mucosa and muscular layer together or overlapping the muscular vesical wall which has been deprived of mucosa (7).
   The upper extremity of this tube was made bulbous to facilitate the anastomosis of the blind proximal end of the sigmoid conduit to the cranial extremity of the bladder tube.
   When osteotomy was performed, the pubic bones were proximated with strong chromic catgut sutures. Since 1975, we avoided osteotomy, and coverage of the closed bladder was provided by groin flaps that also allowed approximation of the hairy skin in the mid line.

Table 2. Bladder exstrophy—colocystoplasty (Arap procedure)

1969–1996	no.	Age	Gender
Cases	56	1y2mo-33y	13 female 14 male
Complete program	27		
Ureterosigmoidostomy	14		
Lost follow-up	11		
Colonic bladder	02		
Death	02		

d. In two different procedures separated by one or two weeks the colonic conduit was anastomosed to the cranial extremity of the bladder tube and the closure of the sigmoid stoma was done later to allow micturition through the urethra (7).

RESULTS

Since 1968, 56 patients with bladder exstrophy were submitted to this program; nowadays 27 patients remain with this procedure; 14 of them had diversion of the colonic bladder to the intestinal tract, 11 were lost to the follow-up; two stayed with colonic conduits and 2 deaths occurred (Table 2).

Continence

Concerning 27 patients reconstructed, the dry intervals are listed in Table 3.

Urinary Diversion

14 patients that had finished the program or not were diverted (transcolonic ureterosigmoidostomy) by several reasons (Table 4).

Ureterocolic Junction Complications

Results from radiologic studies are shown in Table 5.

Late Complications

Late complications occured as shown in Table 6.

Table 3. Bladder exstrophy colocystoplasty continence evaluation

	Dry intervals	No.
16 continent	1 to 2 hs	07
	2hs to 4 hs	06
	C.I.C.*	03
11 incontinents		

C I C= Intermitent catheterism

Table 4. Bladder exstrophy colocystoplasty transcolonic ureterosigmoidostomy

Etiology of diversion	No.
Incontinence	07 (26,3%)
Urethro-trigonal stenosis	05 (18,5%)
Tube dehiscence	01 (3,7%)
Unfavorable bladder	01 (3,7%)
Total	14

Table 5. Bladder exstrophy colocystoplasty ureterocolic junction complications

Ureteral obstruction	17,1% Units*
Reflux	7,2% Units
Total	24,3%

* 4 patients = nephrectomy.

DISCUSSION

The colocystoplasty was performed to avoid frequent complications that occurred in other reconstruction procedures.

The Arap procedure allows reconstruction of even small bladders, utilizing only a detrusor tube to provide continence for an antireflux colocystoplasty. However, in several cases, despite a long detrusor tube previously performed, a large bladder capacity developed. It became clear to us that we were over-utilizing colocystoplasty, and since then we developed in parallel a staged functional reconstruction program.

Regarding continence the children can cooperate to void appropriately; they were trained to void by Valsalva's maneuver. Dry intervals of continence more than 2 hours are possible and usually they use diapers for total protection. The continent patients have normal social life; they study and can perform usual activities; the follow-up is made yearly upon radiologic and ultrasonographic studies and blood chemistry. As we did not use detubularized colonic segments, some patients had minor incontinence related to intestinal peristalsis.

Since 1987, we are routinely detubularizing the intestinal segments.

Complications of the ureterocolic anastomosis are listed in Table 6; there were 14 ureteral units with obstruction, four of them bilateral. Twelve ureters were re-anastomosed at the colonic conduit: 5 with satisfactory results, 3 with residual hydronephrosis and 4 nephrectomies. Reflux was detected in eight units: four submitted to surgical correction,

Table 6. Bladder exstrophy colocystoplasty late complications

Complications	no. Pte
Lithiasis	05
Metabolic disturbances	05
Acute pyelonephritis	03

and four clinically observed; it has been generally well tolerated. Lithiasis was a complication observed in 5 patients; ureteral in two; staghorn in one and 3 in the colonic bladder. The successful antireflux ureterocolic anastomosis is, however, very efficient, even after long term follow up, and extreme vesical distention.

Metabolic disturbances (hyperchloremic hypokalemic acidosis) were observed in five patients: four of them were from the group with ureterosigmoidostomy and one with colocystoplasty. Episodes of high fever, pyelonephritis symptomatology and metabolic acidosis occurred sporadically, specially associated with ureteral obstruction or reflux.

Colocystoplasty was associated with an increased morbidity, specially obstruction and reflux at the ureterocolic anastomosis (4 nephrectomies). Today, reconstruction of bladder exstrophy follows the staged reconstruction program and colocystoplasty is still used in selected cases and not as a routine. We are also detubularizing routinely the intestine for purposes of bladder augmentation.

Bladder augmentation (Arap Procedure) is utilized only for extremes cases, whenever the original bladder is totally inadequate for reconstruction. We also learned that bladder initially very small (after a initial closure without continence) may enlarge and become suitable for anti-incontinence procedure.

REFERENCES

1. Marshall, V.f.; Muecke, E.c. Functional Closure Of Typical Exstrophy Of The Bladder. *J. Urol.*, 104:205–12, 1970.
2. Arap, S.; Gonçalves, E.s.; Gutierrez, E.g.; Góes, G.m. Tratamento Da Extrofia Da Bexiga. I. Reconstrução Plástica. Rev. Assoc. Med. Bras., 22:61–64, 1976.
3. Campos Freire, J.g. Bricker Operation In Children. *Urol. Int.*, 23: 89–93, 1968.
4. Spence, H.m. Ureterosigmoidostomy For Exstrophy Of The Bladder: Results In A Personal Series Of Thirty-one Cases. *Br. J. Urol.*, 38:36–43, 1966.
5. Swenson, O.; Moussatos, G.h.; Fisher, J.h. Results Of Repair Of Exstrophy Of The Bladder. *Surg. Clin. North Am.*, 43:151–5, 1963.
6. Jeffs, R.d.; Guide, S.l.; Oesch, I. The Factors In Successful Exstrophy Closure. *J.urol.*, 127: 974–6, 1982.
7. Arap, S.; Giron, A.m.; Góes, G.m. Complete Reconstruction Of Bladder Exstrophy. *Urology*, 7:413–6, 1976.
8. Young, H.m. Exstrophy Of The Bladder. The First Case In Which A Normal Bladder And Urinary Control Have Been Obtained By Plastic Operations. *Surg. Gynecol. Obstet.*, 74:729–37, 1942.
9. Dees, J.e.; Durham, N.c. Epispadias With Incontinence In The Male. *Surgery*, 12: 621–30, 1942.
10. Leadbetter, G.w. Jr. Surgical Correction Of Total Urinary Incontinence. *J. Urol.*, 91: 261–6, 1964.
11. Couvelaire, R.; Moulonguet, A.; Sagot, A. Sur Le Traitment De L'exstrophie Vésicale. *J. Urol. Nephrol.*, 65: 757–82, 1961.
12. Martinez-piñero, J.a. Tubulation Trigonal: Técnica Quirúrgica Correctora De Incontinencias Urinarias. *Arch. Esp. Urol.*, 21: 166–82, 1968.

19

PENILE LENGTH IN ADULTHOOD AFTER EXSTROPHY RECONSTRUCTION

Richard I. Silver,[2] Andrew Yang,[1] Jacob Ben-Chaim,[1] Robert D. Jeffs,[1] and John P. Gearhart[1]

[1]The Division of Pediatric Urology
Department of Urology
The James Buchanan Brady Urological Institute, and
[2]The Russell H. Morgan Department of Radiology
The Johns Hopkins Hospital
Baltimore, Maryland

ABSTRACT

Purpose

To determine if penile length in men after exstrophy reconstruction is short because of a congenital defect in the size of the corpora cavernosa.

Materials and Methods

Pelvic MRI scans of ten men who underwent exstrophy reconstruction as children and ten age and race matched controls were performed and measurements of penile and pelvic anatomy were compared.

Results

The length of the corpora cavernosa in men after exstrophy reconstruction is shorter than normal. Dividing total corporal length into an anterior and posterior segment reveals that the anterior segment is short but that the posterior segment, attached to the pubic ramus, is normal. However, the diameter of the posterior corporal segment is greater than in controls. Although diastasis of the symphysis pubis increases the intersymphyseal and intercorporal distances, the angle between the corpora cavernosa is unchanged, presumably because the corporal bodies are separated in a parallel fashion.

The Exstrophy–Epispadias Complex, edited by Gearhart and Mathews.
Kluwer Academic / Plenum Publishers, New York, 1999.

Conclusions

The penis in men after exstrophy reconstruction is short, at least in part due to a congenital deficiency of corporal tissue. Since diastasis of the pubic symphysis and chordee reduce penile visibility, approximation of the pubic symphysis and procedures to straighten the penis may improve its cosmetic appearance. However, because the corpora cavernosa are short, penile length in men after exstrophy reconstruction will always be shorter than normal.

INTRODUCTION

Surgical reconstruction of a child born with bladder exstrophy involves both bladder and pelvic closure to repair the ventral body wall defect, bladder neck reconstruction to achieve urinary continence, and epispadias repair to allow direction of the urinary stream. In the male, it also involves techniques to straighten and, thereby, lengthen the penis. The observation that the penis in exstrophy patients who have undergone reconstruction appears short is a problem frustrating to patients that must live with it and to surgeons that have attempted to repair it. However, the cause of this problem remains unclear.

Two reasons have been identified to explain why the penis in men after exstrophy reconstruction might look short. First, diastasis of the pubic symphysis may separate the corporal bodies and draw penile tissue into the pelvis, creating the illusion of a short penis despite corpora cavernosa that are normal in length.[1,2] Second, independent of the surrounding pelvic anatomy, the corpora cavernosa may themselves be congenitally short.[3] Of course, both anatomic abnormalities may coexist and contribute to this problem. Since accurate measurements of the penis in men with exstrophy have not yet been performed, this study was designed to determine if the length of the corpora cavernosa is a factor in the results of exstrophy reconstruction.

MATERIALS AND METHODS

Magnetic resonance imaging (MRI) scans of the pelvis of ten men (ages 17–38) born with exstrophy who underwent reconstruction as children were reviewed and compared to an age and race matched case control set of pelvic MRI scans of normal men (ages 29–40). Measurements were obtained for the following: the intersymphyseal distance (ISD) (Figures 1 and 2); the intercorporal distance at the attachment of the penis to the pelvis (ICD) (Figures 1 and 2); corporal diameter (Cdiam) (Figures 1 and 2); the angle subtended by the corpora cavernosa (aCC) (Figures 3 and 4); the angle subtended by the puborectalis muscle (aPRM) (Figures 3 and 4); and total corporal length (TCL). TCL was determined by measuring it in two components: an anterior segment (ACL) that was ventral to the pelvic attachment (Figures 5 and 6) and a posterior segment (PCL) that was attached to and supported by the bony pelvis (Figures 7 and 8). The sum of the ACL and PCL provided the TCL. Means and standard deviations were calculated and compared by a two-tailed unpaired t test assuming unequal variance, with significance considered for p values less than 0.05.

RESULTS

The data obtained by inspection of the pelvic MRI scans are presented in Table 1, and the manner in which they were obtained is presented in the accompanying figure. No

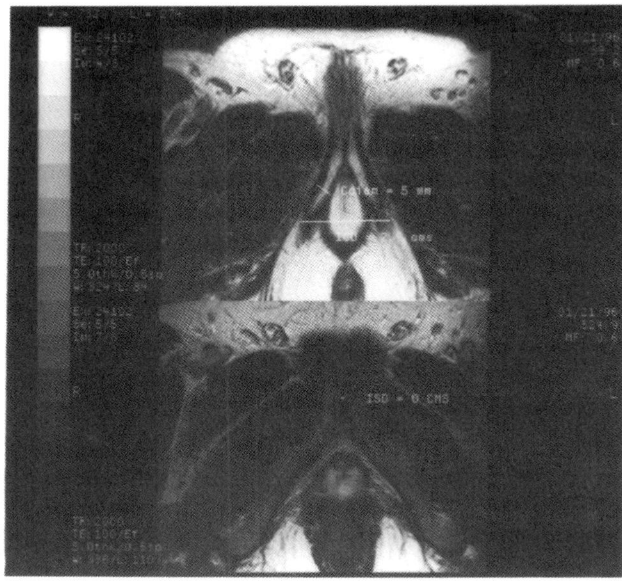

Figure 1. MRI scan of normal adult male pelvis, axial view. The diameter of the corpus cavernosum (Cdiam) is 5 mm, the intercorporal distance (ICD) is 2.9 cm, and the intersymphyseal distance (ISD) is 0 cm.

statistical comparison was performed for the intersymphyseal distance because pubic symphysis diastasis is part of the exstrophy complex and a difference exists by definition. Measurements that are significantly different between the control and exstrophy groups include the intercorporal distance, the angle of the puborectalis muscle, corporal diameter, total corporal length, and anterior corporal length. Interestingly, the angle between the corpora cavernosa in these patients is not statistically different from controls. Theoretically, this may be because the pubic symphysis diastasis in exstrophy increases both the intersymphyseal and intercorporal distances to a similar degree, and, therefore, does not significantly alter the angle between the corpora cavernosa. Another interesting and unexpected finding is that the posterior corporal length is not statistically different between these groups.

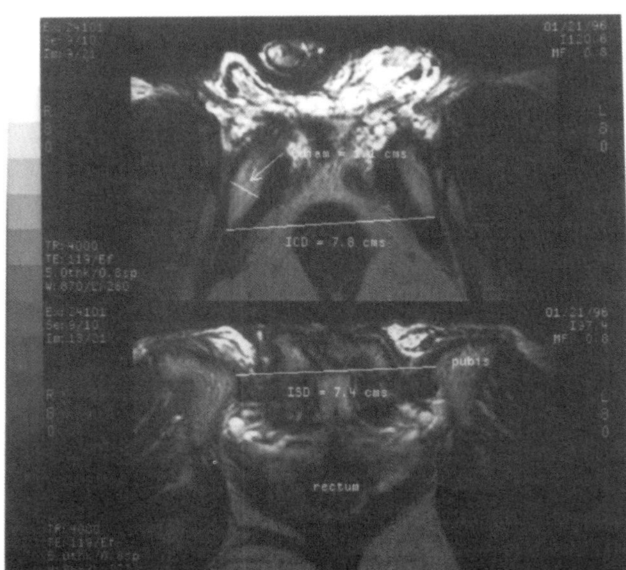

Figure 2. MRI scan of exstrophy adult male pelvis, axial view. The diameter of the corpus cavernosum (Cdiam) is 1.1 cm, the intercorporal distance (ICD) is 7.8 cm, and the intersymphyseal distance (ISD) is 7.4 cm.

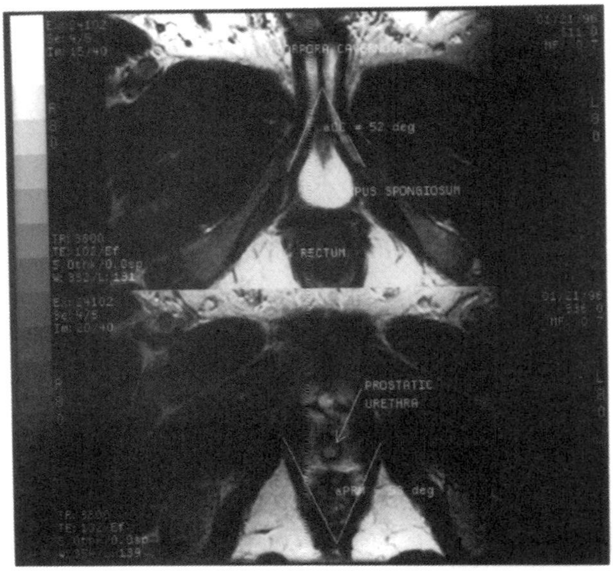

Figure 3. MRI scan of normal adult male pelvis, axial view. The angle subtended by the corpora cavernosa (aCC) is 52° and the angle subtended by the puborectalis muscle (aPRM) is 53°.

DISCUSSION

The embryology of exstrophy causes severe deformity to almost all aspects of the lower genitourinary tract, including the penis. The exstrophy penis is shortened and flattened with a splayed glans and a ventral prepuce. Complete epispadias is present with an exposed urethra. The penis is displaced cephalad with severe dorsal chordee due to insufficient skin, fibrous bands in the penopubic region, and a short urethra.[1] The corpora cavernosa, which may be asymmetric, remain separate throughout the length of the shaft and the nerve supply enters laterally.[4] There is pubic symphysis diastasis and the pubic rami are externally rotated and shortened by 30%.[5] Experts generally agree that some corporal

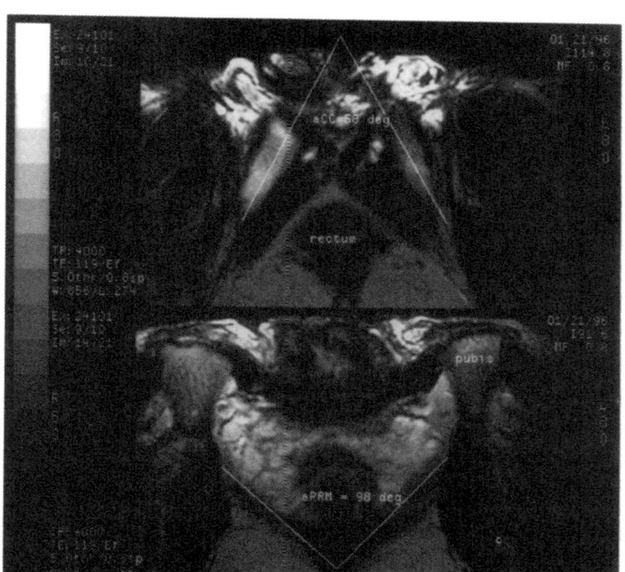

Figure 4. MRI scan of exstrophy adult male pelvis, axial view. The angle subtended by the corpora cavernosa (aCC) is 58° and the angle subtended by the puborectalis muscle (aPRM) is 98°.

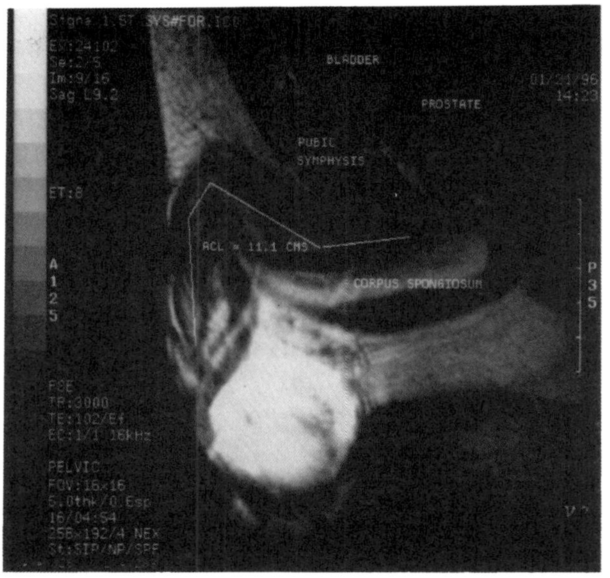

Figure 5. MRI scan of normal adult male pelvis, sagittal view. The anterior corporal length (ACL) is 11.1 cm.

length is hidden because of the pubic symphysis diastasis, which separates the corpora cavernosa and draws some corporal tissue into the pelvis, hiding it between the pubic symphysis.

For the male with exstrophy, many procedures to straighten the penis to make it look longer have been described.[2,6–17] These include techniques to release the dorsal chordee, lengthen the urethra, and provide skin coverage. Most experts also advocate partial mobilization of the corporal bodies from the pubic rami, while leaving the posterior portion attached. Others, such as Kelly and Eraklis, have proposed total detachment of the corporal bodies from the pubis.[17] However, most surgeons do not favor this radical approach because it risks total denervation and devascularization of the penis.

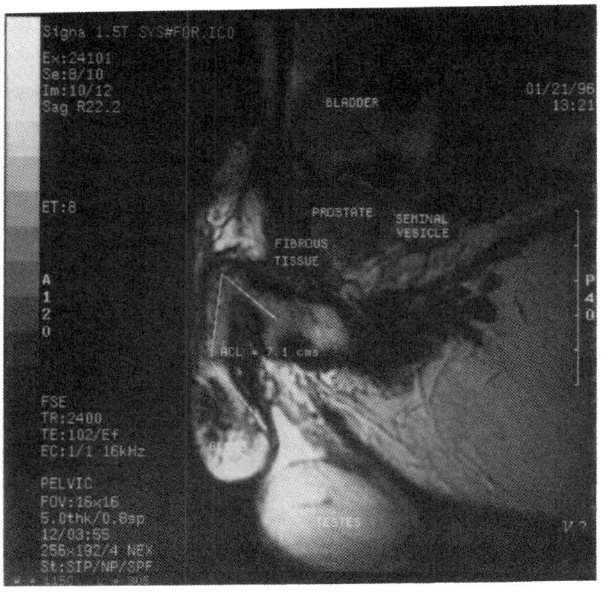

Figure 6. MRI scan of exstrophy adult male pelvis, sagittal view. The anterior corporal length (ACL) is 7.1 cm.

Table 1. Comparison of penile and pelvic measurements in adult male controls and exstrophy patients

Abbreviation	Measurement	Control	Exstrophy	P value
ISD	intersymphyseal distance	0 cm	7.9±1.9 cm	N/A
ICD	intercorporal distance	4.5±0.5 cm	6.5±2.2 cm	<.02
aCC	angle between corpora cavernosa	50±9°	64±24°	=0.10
aPRM	angle of puborectalis muscle	52±15°	86±19°	<.001
Cdiam	corporal diameter	1.0±0.2 cm	1.4±0.4 cm	<.02
TCL	total corporal length	16.2±2.4 cm	10.1±2.5 cm	<.0001
-ACL	anterior corporal length	12.3±2.2 cm	6.9±2.4 cm	<.0001
-PCL	posterior corporal length	3.9±1.2 cm	3.2±1.4 cm	=0.28

At this institution, neonatal males with exstrophy usually undergo initial penile straightening at the time of primary bladder closure. This is accomplished by dividing all tethering attachments to the penis, including the skin and suspensory ligaments, and detaching the urogenital diaphragm from the body of the pubis. Division of the urethra distal to the verumontanum and paraexstrophy skin flaps are used when necessary to lengthen the urethra.[18,19] Limited mobilization of the corpora cavernosa, with care to preserve the neurovascular supply, and approximation of the pubic rami, usually with anterior innominate and iliac osteotomy,[20] help to contribute to the cosmetic appearance after the first procedure. At the age of one to two years a Cantwell-Ransley epispadias repair is usually performed, which closes the urethra and transfers it ventrally; simultaneously, the dorsal chordee is corrected by rotation of the corporal bodies over the neourethra.[21] Even after an aggressive staged reconstruction, however, the exstrophy penis almost always looks shorter than age matched controls.

Some experts believe that the length of the penis in exstrophy is normal but that the deformity of the bony pelvis causes the penis to look short. In 1975, Johnston stated that "the shortness of the penis is caused mainly by the separation of the pubic bones. The corpora cavernosa are, themselves, of normal length but much of the length is wasted between the shaft and the bony attachment."[1] In 1991, McLorie et al., used an *in vitro* model

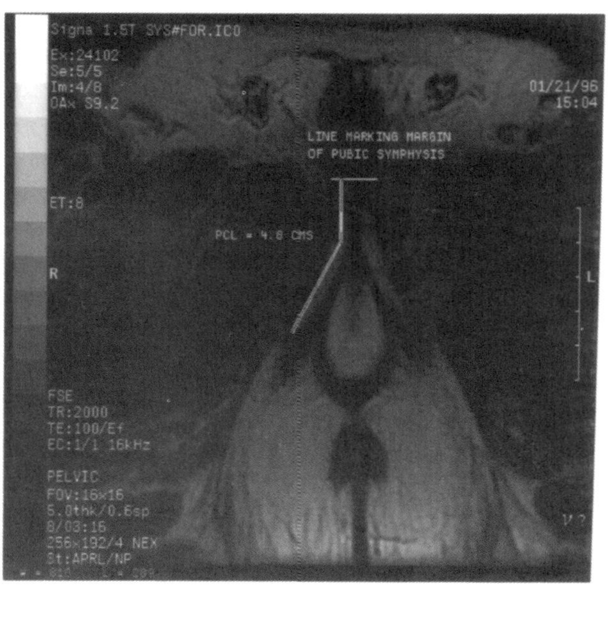

Figure 7. MRI scan of normal adult male pelvis, axial view. The posterior corporal length (PCL) is 4.8 cm.

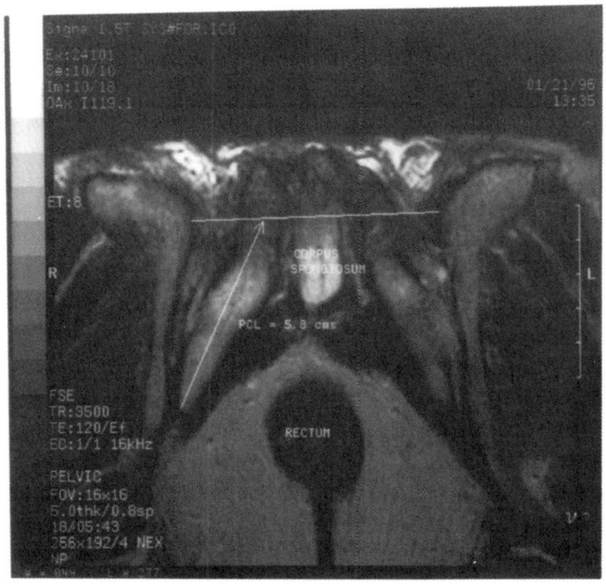

Figure 8. MRI scan of exstrophy adult male pelvis, axial view. The posterior corporal length (PCL) is 5.8 cm.

of the exstrophy pelvis to prove that simply closing the pubic symphysis diastasis can make the penis appear longer.[22] However, since data were not available regarding the length of the corpora cavernosa in exstrophy, an inflatable penile prosthesis was used to represent the penis in this model.

In 1984 a study by Woodhouse and Kellett provided the first experimental evidence that the corpora cavernosa in exstrophy might be shorter than normal. They used cavernosography to examine penile deformities in 17 adult patients with exstrophy and epispadias. Findings included that the deep part of the corpora cavernosa are of "normal caliber;" that the "exophytic," or visible part, of the penis is short; and that total penile length in exstrophy is adequate for intercourse, but that it is shorter than in normal men.[3] Although this technique may have been the best available at that time, and these statements may have been true, accurate measurements of penile length were not performed. Therefore, the extent of the difference in penile length between men who are normal and those who have exstrophy was not determined, obviating statistical interpretation.

The use of magnetic resonance imaging in this study to examine penile length in exstrophy has some advantages over previously used techniques. It allows for detailed and accurate measurements of both the penis and surrounding anatomy, is non-invasive, and does not expose subjects to radiation or contrast dye. The use of three dimensional MRI image reconstruction is another approach that may be helpful in this regard, and which is being considered for further study of exstrophy anatomy.

Using MRI, it is possible to measure the length of the corpus cavernosum in two segments: (1) an anterior segment ventral to the pelvic skeleton that is visible to the patient and other observers, and (2) a posterior segment that is attached to the pubic ramus and is hidden within the pelvis. This delineation is important because, by these definitions, anterior corporal length is what really matters to the patient because it determines his appearance and his function during sexual intercourse. In contrast, the posterior segment anchors the penis to the pelvis and surrounds its neurovascular supply, but is of little clinical concern to the patient. Interestingly, the results of this study show that anterior corporal length is short in exstrophy, but that the posterior corporal length is normal. As a result, total corporal length is also short because of the contribution of the anterior component. Of additional interest is the finding that, although the posterior corporal segment has a normal length, it also has a greater diameter than that of normal men. This does not seem to be due to post-operative scarring since this portion of the penis is not usually part of the dissection in exstrophy reconstruction, including at this institution.

Unfortunately, these findings suggest that there is little room for improvement in current surgical results. Penile "lengthening" procedures improve apparent penile length because they straighten the penis, not by transfering tissue to the corporal bodies. Moreover, since the length of the posterior corporal segment, the portion that is attached to the pubic ramus, is normal in exstrophy, the benefit of aggressive mobilization and detachment does not seem worth the risks of denervation and devascularization

Pelvic skeletal anatomy in exstrophy patients, even after reconstruction, also remains abnormal. Despite osteotomy and surgical approximation of the pubic symphysis in infants, over time the pubic bones tends to drift apart and separate.[22,23] This is reflected in the increased intersymphyseal distance, intercorporal distance, and puborectalis muscle angle seen in this study. However, it is a surprise to find that the geometric angle subtended by the corpora cavernosa is not significantly different from that of normal men. Apparently the bony deformity of the pelvis, including the diastasis of the pubic symphysis, has a minimal effect on this relationship.

It is clear that with this experimental approach an erection during MRI scanning could alter penile dimensions and lead to inaccurate data. However, there was no evidence of an erect penis on any of the MRI scans used in this study.

The self-esteem of adolescent boys and young men depends to a great extent on their body image, and certainly the appearance of their external genitalia is an important component. In fact, male adolescents consider the abnormal appearance of their external genitalia to be a greater psychosocial problem for them than incontinence.[24] As adults, however, most are satisfied with their cosmetic results.[25–29] All have normal libido and most have adequate penile length for sexual intercourse. Some even father children.[27–29] For these reasons, the finding that penile length in exstrophy is congenitally short is not intended to be an argument for sex reassignment of a newborn male with exstrophy. Rather, these findings should help both the exstrophy patient and surgeon understand the extent of the penile deformity and why surgery can improve, but not completely correct, the condition.

CONCLUSIONS

Penile length in men after exstrophy reconstruction is shorter than normal because of a congenitally short anterior corporal segment. The length of the posterior corporal segment is normal, although its diameter is wider than that of controls. Therefore, although it may be valid that diastasis of the pubic symphysis can make the exstrophy penis look short, the penis itself is short because of a congenital deficiency of corporal tissue. This implies that even after an ideal reconstruction, including complete closure of the pubic symphysis diastasis and straightening of the penis, the man born with exstrophy will still have a penis that is shorter than normal.

ACKNOWLEDGMENT

The assistance of Dr. H. Ballentine Carter in retrieving the control set data is deeply appreciated.

PUBLISHED REFERENCE

Journal of Urology 157: 999–1003, 1997.

REFERENCES

1. Johnston, JH. The Genital Aspects of Exstrophy. J Urol 113:701, 1975.
2. Schillinger, JF and Wiley, MJ. Bladder Exstrophy: Penile Lengthening Procedure. Urology 24:434, 1984.
3. Woodhouse, CRJ and Kellett, MJ. Anatomy of the Penis and Its Deformities in Exstrophy and Epispadias. J Urol 132:1122, 1984.
4. Hurwitz, RS, Woodhouse, CRJ, and Ransley, P. The Anatomical Course of the Neurovascular Bundles in Epispadias. J Urol 136:68, 1986.
5. Sponseller, PD, Bisson, LJ, Gearhart, JP, Jeffs, RD, Magid, D, and Fishman, E. The Anatomy of the Pelvis in the Exstrophy Complex. J Bone Joint Surg 77A:177, 1995.
6. Hinman, Jr., F. A Method of Lengthening and Repairing the Penis in Exstrophy of the Bladder. J Urol 79:237, 1958.
7. Hendren, W. Hardy. Penile Lengthening After Previous Repair of Epispadias. J Urol 121:527, 1979.
8. Tanagho, E. Male Epispadias: Surgical Repairs of Urethropenile Deformity. Br J Urol 48:127, 1976.
9. Johnston, JH. Lengthening of the Congenital or Acquired Short Penis. Br J Urol 46:685, 1974.
10. Allen, TD, Spence, HM, and Salyer, KE. Reconstruction of the External Genitalia in Exstrophy of the Bladder: Preliminary Communication. J Urol 111:830, 1974.
11. Vorstman, B, Horton, CE, Winslow, BH. Repair of Secondary Genital Deformities of Epispadias/Exstrophy. Clinics in Plastic Surgery 15:381, 1988.
12. Woodhouse, CRJ. Reconstruction of the Epispadiac Penis in Adolescents. Progress in Pediatric Surgery 23:165, 1989.
13. Koff, SA and Eakins, M. The Treatment of Penile Chordee Using Corporeal Rotation. J Urol 131:931, 1984.
14. Johnston, JH and Kogan, SJ. The Exstrophic Anomalies and Their Surgical Reconstruction. Curr Probs Surg:1–39, August 1974.
15. Williams, DI. Epispadias and Exstrophy. Proc Roy Soc Med 62:1079–1081, 1969.
16. Michalowski, M and Modelski, W. The Surgical Treatment of Epispadias. SGO 117:465, 1963.
17. Kelley, JH and Eraklis, AJ. A Procedure for Lengthening the Phallus in Boys with Exstrophy of the Bladder. J Ped Surg 6:645, 1971.
18. Lepor, H, Shapiro, E, and Jeffs, RD. Urethral Reconstruction in Boys with Classical Bladder Exstrophy. J Urol 131:512, 1984.
19. Duckett, JW. Use of Paraexstrophy Skin Pedicle Grafts for Correction of Exstrophy and Epispadias Repair. Birth Defects 13:171, 1977.
20. Sponseller, PD. Anterior Pelvic Osteotomy of Superior Pubic Ramus to Facilitate Bladder Exstrophy Repair, in Dialogues in Pediatric Urology, Osteotomy and the Bladder Exstrophy Patient 16 (1):3, 1993.
21. Gearhart, JP, Leonard, MP, Burgers, JK, and Jeffs, RD. The Cantwell-Ransley Technique for Repair of Epispadias. J Urol 148:851, 1992.
22. McLorie, GA, Bellemore, MC, and Salter, RB. Penile Deformity in Bladder Exstrophy: Correlation With Closure of Pelvic Defect. J Ped Surg 26:201, 1991.
23. Gearhart, JP, Forschner, DC, Jeffs, RD, and Sponseller, PD. A Combined Vertical and Horizontal Pelvic Osteotomy Approach for the Initial and Secondary Repair of Bladder Exstrophy. J Urol 155:689, 1996.
24. Feinberg, T, Lattimer, JK, Jeter, K, Langford, W, and Beck, L. Questions That Worry Children with Exstrophy. Pediatrics 53:242, 1974.
25. Mesrobian, HJ, Kelalis, PP, and Kramer, SA. Long-Term Followup of Cosmetic Appearance and Genital Function in Boys with Exstrophy: Review of 53 Patients. J Urol 136:256, 1986.
26. Woodhouse, CRJ, Ransley, PG, and Williams, DI. The Patient With Exstrophy in Adult Life. Br J Urol 55:632, 1983.
27. Hanna, MK and Williams, DI. Genital Function in Males with Vesical Exstrophy and Epispadias. Br J Urol 44:169, 1972.
28. Woodhouse, CRJ. The Sexual and Reproductive Consequences of Congenital Genitourinary Anomalies. J Urol 152:645, 1994.
29. Lattimer, JK, Beck, L, Yeaw, S, Puchner, PJ, Macfarlane, MT, and Krisiloff, M. Long-Term Followup After Exstrophy Closure: Late Improvement and Good Quality of Life. J Urol 119:664, 1978.

20

COMPLETE PENILE DISASSEMBLY FOR EPISPADIAS REPAIR

Michael E. Mitchell

Children's Hospital and Medical Center
Seattle, Washington

ABSTRACT

Over the past six years we have used a technique for epispadias repair which is based on the unique anatomy of the epispadias deformity which is actually a deformation rather than a malformation (e.g. as in hypospadias). In this technique the unique and separate blood supply of the urethral plate, the glans penis and the corpora cavernosa facilitate complete disassembly of the penis which enables reassembly resulting in normal anatomic position of the urethra and bladder neck. This repair also enables complete and independent rotation of the corpora cavernosa to better correct the dorsal deflection deformity.

This technique has made possible reconstruction in staged exstrophy closures, but more importantly has realized the potential for *primary complete reconstruction of the male exstrophy patient*. It can also be applicable to redo reconstructions in epispadias patients who have had multiple procedures as it facilitates reconstruction with flap urethroplasties or grafts (buccal mucosa).

This procedure facilitates the tubularization and ventralization of the entire distal urethra/complete glans repair and results in a normal appearing phallus. In primary repair patients the resultant urethra is straight and easily catheterized.

INTRODUCTION

In Cantwell's first description of epispadias, almost a hundred years ago, he described the repair of epispadias and considered that epispadias was a deformation.[1] In his repair he recommended complete disassembly of the penis, however, in actuality he did not completely do this because he failed to "deglove" penile skin from the ventrum of the phallus. Over the years the repair of the epispadias phallus changed, but in general the anatomic and functional outcome was marginal.[2] Although this initial rather aggressive

The Exstrophy–Epispadias Complex, edited by Gearhart and Mathews.
Kluwer Academic / Plenum Publishers, New York, 1999.

description of Cantwell came early, later surgeons felt that such a repair was associated with unacceptably high rates of fistula formation and more conservative approaches to urethral reconstruction were described.[3] Subsequently, many different descriptions of urethral reconstruction in the epispadias repair have been described. Many were somewhat limited by the fact that the penis was not totally dissected. Multi-staged procedures and grafting for urethra were also proposed.

In this complete penile disassembly technique the unique blood supply of the penis in epispadias is utilized. The blood supply of the corpora cavernosa comes through the right and left central arteries. There is very little cross circulation between the corporal bodies. The glans penis also can be separated because each glans half is associated with its independent blood supply coming in the dorsal neurovascular bundle which in the epispadias penis is on the lateral aspect of the laterally rotated corpora cavernosa. The urethral plate has independent blood supply from the prostatic urethra coursing through the corpora spongeosa which is usually very well developed and forms a triangle of tissue ventral to the urethra plate that goes between the corpora cavernosa. The urethral plate with the corpora spongeosa can be carefully dissected *completely* off of the corpora cavernosa and also *separated* from the glans penis as very little distal circulation and collateral circulation come from the distal glans to the proximal urethra. Each corpora cavernosa with hemiglans can be separated such that at the completion of the dissection there is a urethral plate with a ventral corpora spongeosa and a right and a left corpora cavernosa with hemiglans. After complete disassembly the penis can be reconstructed tubularizing the urethral plate bringing the corporal cavernosa together in the midline and rotating medially such that the lateral dorsal complex is rotated to the dorsum. The glans can be reconfigured and the tubularized urethral plate can be brought along the ventral groove of the approximated corpora cavernosa. Usually the plate is long enough to go right out the tip of the reconstructed glans.

SURGICAL TECHNIQUE

Figure 1 shows the anatomy of the epispadias with dorsal deflection, open urethral plate and glans penis with separation of the corpora cavernosa. Figure 2: The dissection is along the lateral aspects of the urethral plate *and on the ventrum of the penis*. Care is taken to avoid the dorsal complex which is along the lateral aspect of the phallus (Figure 2a). Buck's fascia can be seen on the lateral and ventral aspect of the penis, but terminates at the urethral plate. As shown in Figure 2b the urethral plate is dissected off of the corpora cavernosa and the dissection plane is on the tunica albuginea of the corpora cavernosa. This dissection is relatively bloodless and can be extended right down to the tip of the corpora cavernosa. Note the configuration of the urethral plate is not that of a plate, but ventral triangle with a very healthy wedge of spongeosa tissue. Once the urethral plate is dissected off of the corpora (Figure 3a) the corpora can be easily separated. This separation and the complete urethral plate dissection is facilitated by the initial ventral dissection of the skin off of the corpora cavernosa (Figure 3b).

The penis is now completely disassembled, but the dissection must be carried proximally along the corpora cavernosa right to the base of the bladder and around the bladder neck as shown in Figures 4a and 4b. At this point the bladder can be closed if this is an exstrophy patient or previously poorly closed exstrophy patient. in the pure epispadias patient, of course, the urethra will be closed only to the area of the proximal urethra. Closure of the bladder is shown in Figure 4a and of the urethra in 4b (we usually use a running 5–0 or 6–0 Vicryl suture for this purpose). The tubularized urethral plate is placed on the ven-

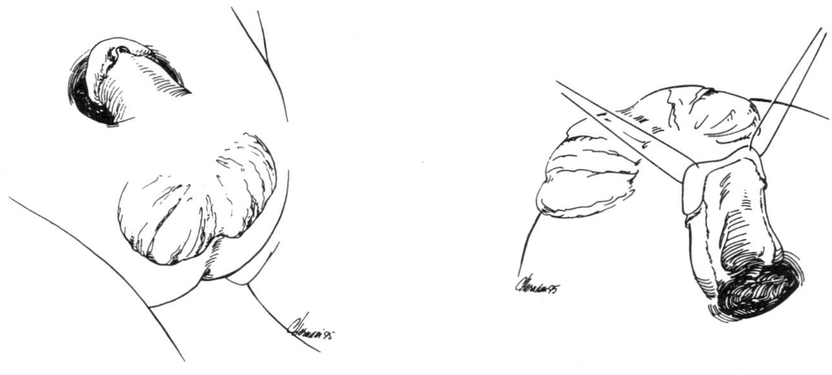

A B

Figure 1. A, B. The anatomy of the epispadias shows a dorsal deflection with minimal chordee. The dorsal bending usually is dependent on the previous surgical procedures formed rather than the original pathology. This patient previously had a partially failed exstrophy closure.

trum of the phallus after medial rotation of the corporal bodies as shown in Figure 5a and 5b. The glans is reconstructed as shown in Figure 5B. Note that it is usually necessary to resect a wedge of glanular tissue on the dorsum of the glans as the glans tissue is often redundant in this area. The penile skin is then wrapped around the newly reconstructed phallus and dorsal closure is performed preferably with a Z-plasty to prevent scar contracture and dorsal shortening (Figure 5c).

RESULTS

We reported our initial results with this repair in January of 1996 in *The Journal of Urology*, Volume 155, pages 300–304. As of that time we had used the repair in ten patients. We have subsequently more than doubled this experience and have been very

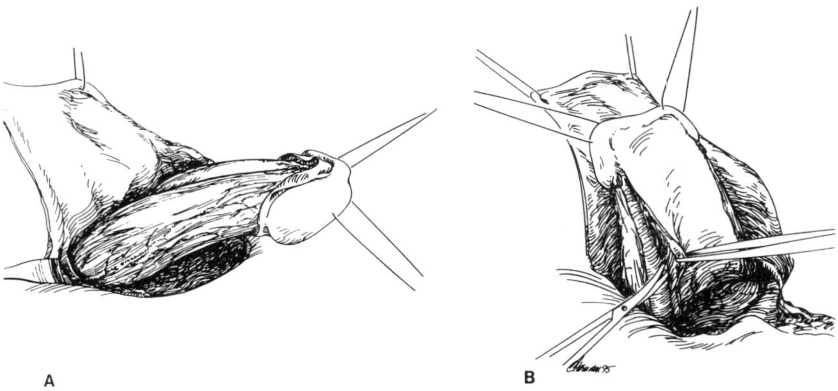

A B

Figure 2. A. A dissection is initiated along the urethral plate and then the skin of the ventrum of the penis is dissected off of Buck's fascia. B. The urethral plate is then dissected off the corpora cavernosa. The spongeosa must be preserved with urethral plate; the plane of dissection is on the tunica albuginea of the corpora cavernosa.

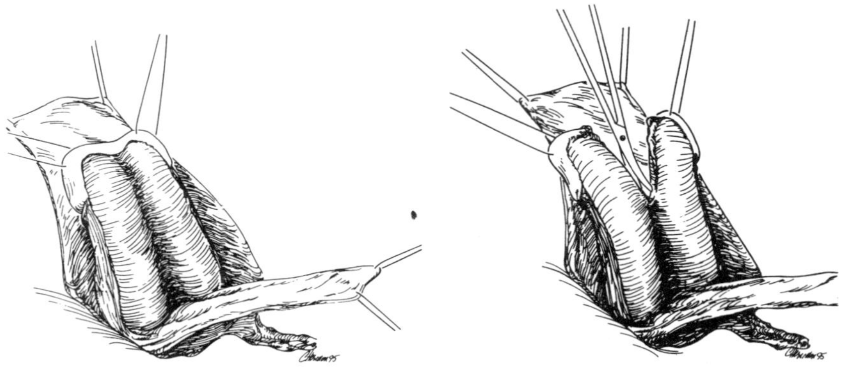

A B

Figure 3. A. The urethral plate has been dissected, but note that it is actually more of a triangular-shaped flap with a flat dorsum. B. The separation of the corpora cavernosa can be done with impunity as there is very little cross circulation. The hemiglans caps are kept with each corporal body. This dissection can be extended very far proximally on the medial aspect of the corpora as this plane is a very safe one.

pleased with this technique. The advantages are the degree of freedom in reconstruction which enables the construction of a more cosmetically appealing phallus with straight erections. It should be noted that there are two components of the epispadias which result in different dorsal deformities. These are dorsal deflection and dorsal curvature (dorsal chordee). Dorsal curvature has to be repaired by shortening the ventrum of the phallus or excision of scar tissue and dermal grafting or suturing the corpora together after incision of the tunica albuginea. Dorsal deflection is almost completely repaired by simply rotating the two corpora cavernosa medially as described in the repair. It has been my experience that if the epispadias is repaired early, there is minimal dorsal chordee. The majority of the deformity is dorsal deflection.

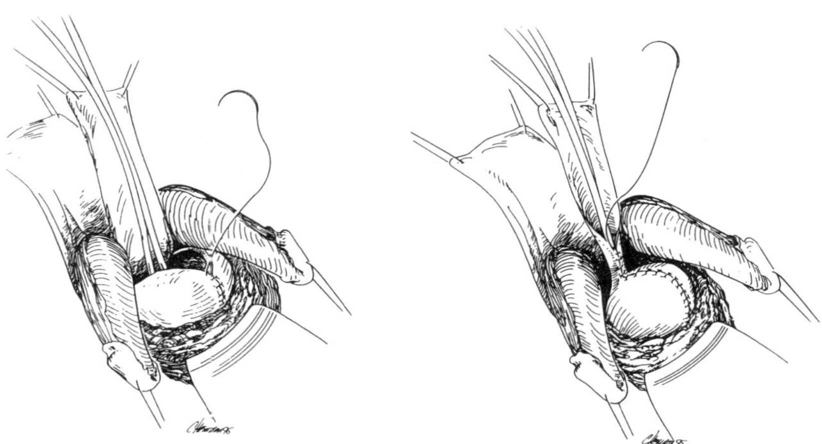

A B

Figure 4. A. Closing of the bladder plate after extensive dissection in the area of the bladder neck on each side. B. Tubularization of the urethra. Note how the bladder neck and urethra are now situated well posterior to the corpora cavernosa.

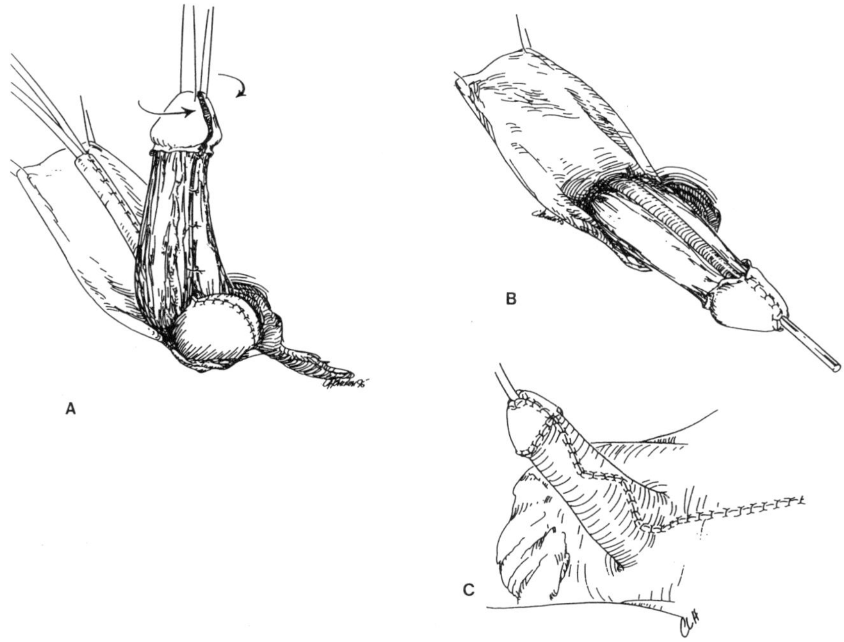

Figure 5. A. The corpora are rotated and sutured in the dorsal midline to maintain this rotation. The tubularized urethra is now on the ventrum of the phallus. B. Glans is complete with the appropriate placement of the urethra. C. The ventral skin is usually ample to cover the penis. Efforts should be made to perform a Z-plasty dorsally. Usually a suture or two at the base of the phallus is important to keep the skin from sliding distally onto the shaft of the reconstructed phallus.

DISCUSSION

The complete penile disassembly has enabled not only the reconstruction of the epispadias, *but facilitated primary bladder closure in boys with penile repair*. The advantage of early penile repair is that bladder neck in the male can be moved posteriorly to facilitate appropriate position in the pelvis. This greatly facilitates primary bladder function (bladder cycling) which facilitates potential for primary continence in the male. In my opinion, *it is the epispadias which dictates bladder exstrophy bladder repair and should be done in the newborn*. We now have experience with six boys who have had complete primary reconstruction and the majority of these boys have primary continence and excellent penile repairs.

This is the report of a new technique which facilitates complete penile repair and appropriate anatomic positioning of the urethra and bladder neck. Using this repair alleviates the need for paravesical flaps which should never be used either in the male or female in bladder closure or reconstruction.

REFERENCES

1. Cantwell, F.V.: Operative treatment of epispadias by transplantation of the urethra. Ann. Surg., 22:689, 1895.
2. Kramer, S.A., Mesrobian, H.-G.J. and Kelalis, P.P.: Long-term follow-up of cosmetic appearance and genital function in male epispadias: review of 70 patients. J. Urol., 135:543, 1986.
3. Young, H.H.: An operation for the cure of incontinence associated with epispadias. J. Urol., 7:1, 1922.

21

THE CANTWELL-RANSLEY EPISPADIAS REPAIR IN EXSTROPHY/EPISPADIAS

Lessons Learned

John P. Gearhart,* Christopher Sciortino, Jacob Ben-Chaim, Dennis S. Peppas, and Robert D. Jeffs

The Johns Hopkins Hospital, and
The Johns Hopkins University School of Medicine
James Buchanan Brady Urological Institute
Department of Urology
The Division of Pediatric Urology
Baltimore, Maryland 21287-2101

ABSTRACT

Objectives

We evaluate our experience with the Cantwell-Ransley epispadias repair, in order to determined the lessons that have been learned with the increased experience and follow-up.

Methods

A total of 75 boys (60 with bladder exstrophy and 15 with complete epispadias) underwent a Cantwell-Ransley Epispadias repair in our institute at the last 6 years. Primary repair was performed in 58 boys (45 with exstrophy and 13 with epispadias) and secondary repair—after prior failed closure in 17 boys (12 at the secondary exstrophy closure, 3 with exstrophy and 3 with complete epispadias).

---

* Contact author: John P. Gearhart, M.D., Marburg 149, The Johns Hopkins Hospital, Baltimore, MD 21287-2101, Fax #: (410) 955 0833, Phone #: (410) 955 8710

The Exstrophy–Epispadias Complex, edited by Gearhart and Mathews.
Kluwer Academic / Plenum Publishers, New York, 1999.

Results

At a mean follow-up of 28 months, all patients had a horizontal or downward angled penis while standing. The incidence of urethrocutaneous fistulae in the immediate postoperative state was 21% and at three months was 15%. The incidence of urethrocutaneous fistulae was no more in those patients in whom paraexstrophy skin flaps were used at anterior closure as in those in whom the urethral plate was left intact. Two patients developed a urethral stricture at the proximal anastomotic area, and four patients had minor skin separation of the dorsal penile skin closure. Catheterization and/or cystoscopy has been performed in 60 patients and revealed an easily negotiable urethral channel in all.

Conclusions

The Cantwell-Ransley epispadias offers a straighter urethra, better correction of chordee and cosmetic result and a lower fistula rate in the exstrophy/epispadias patient.

Formerly, the modified Young repair had been the most commonly used technique in our institution to repair the urethra in classic bladder exstrophy and complete epispadias.[1] However, the urethra was very tortuous, and the fistula rate was unacceptably high (21%).[2] In 1988, the authors began to use the Cantwell-Ransley repair exclusively for the repair of epispadias in patients with classic bladder exstrophy or epispadias and have reported this experience.[3] Recently, the authors have expanded their use of this technique and review their results with 75 Cantwell-Ransley Urethroplasties performed between 1988 and 1994 in the exstrophy/epispadias population.

PATIENTS AND METHODS

The authors studied 75 male patients with either classic bladder exstrophy (60 patients) or epispadias (15 patients) who were referred to their hospital for reconstruction between 1988 and 1994. Patient age ranged from 1.5 years to 18 years with an average age of 2.8 years. Of 60 patients with bladder exstrophy, 26 had a short urethral groove and required paraexstrophy skin flaps for penile lengthening at the time of bladder closure.[4] Of the remaining 15 patients, 12 had penopubic and 3 had penile epispadias at presentation.

This technique was used for primary urethroplasty in 45 patients with classic bladder exstrophy and epispadias in 13 patients. Also, it was combined with reclosure of bladder exstrophy in 12 patients, used as a secondary procedure after an initial failed urethroplasty in 3 exstrophy patients and 2 epispadias patients (Table 1).

The epispadias is repaired when the patient is 2 to 3 years old. The surgical technique is similar to that described in detail in our previous communication.[3] However, a

Table 1. Application of Cantwell-Ransley urethroplasty

Condition	#
Exstrophy	45
Secondary exstrophy closure with epispadias repair	12
Failed prior epispadias repair in exstrophy patients	3
Primary epispadias (penopubic and penile)	13
Failed prior epispadias repair	2

few changes have been made including: (1) complete mobilization of the corpora spongiosum along with the urethral plate; and (2) avoidance of routine mobilization of the neurovascular bundles, since in many cases the incision and the anastomosis or suturing of the corporal bodies can be safely done without their mobilization.

At operation, indwelling urethral silicone stents were placed and were left for 12–14 days postoperatively. In those patients who underwent a reclosure of their bladder exstrophy, along with a Cantwell-Ransley repair, indwelling ureteral stents were left for 2–3 weeks, and a suprapubic tube for 4–6 weeks. In those patients who have had urinary diversion, urethral self catheterization was performed through the newly constructed urethra once a day for 3 months postoperatively.

In addition, in order to find whether the technique of epispadias repair affects the bladder outlet resistance, and the eventual bladder capacity, 20 age matched patients who had undergone a Cantwell-Ransley repair had their bladder capacities pre- and post-epispadias repair compared with bladder capacities in an equal number of patients who had undergone a Young epispadias repair. In all of the patients the bladder capacity was measured under general anesthesia prior and 2 years after epispadias repair. The results of epispadias repair were assessed by physical examination, and by patient and parental interviews during a follow-up clinic visits.

RESULTS

With this technique, follow up has ranged from 6 to 72 months (average of 28 months). Fistulas developed in 16 patients (21%) in the immediate postoperative period: fourteen in patients with classic exstrophy and two with complete epispadias. In four patients the fistulas closed spontaneously in the first three months after surgery (1 epispadias, 2 exstrophy patients with paraexstrophy skin flaps and one exstrophy patient with an intact urethral plate). Twelve patients underwent a separate procedure to close their urethrocutaneous fistulas. All of these fistulas occurred at the base of the penis. Of the fistulas in the exstrophy group, 9 of 49 (18%) occurred in patients with an intact urethral plate and 5 of 26 patients (19%) in whom paraexstrophy flaps were used at the time of initial closure. Two older patients had free full thickness grafts (1 ureter, 1 buccal mucosa) placed and brought under the corpora with an excellent functional and cosmetic appearance after a prior failed epispadias repair. Four patients had minimal separation of the dorsal skin incision which granulated without incident. Two patients with bladder exstrophy who had undergone a prior failed epispadias repair, and in whom a full thickness skin graft was used for their secondary repair, developed a filmy stricture at the proximal anastomosis which responded to direct vision internal urethrotomy. One exstrophy patient who had undergone a prior Young epispadias repair and developed a urethral diverticulum was treated with a Cantwell-Ransley repair with no recurrence of the diverticulum.

The direction of the penis when standing was downward or horizontal in all patients and the appearance of the penis was acceptable to both the parents and children. Seven the 75 patients who are older than 17 years of age, have engaged in satisfactory intercourse and say their penis is quite acceptable functionally and cosmetically. A total of 68 patients either void or drain incontinently per urethra and 2 have been previously diverted. Sixty of these patients have undergone either cystoscopy or urethral catheterization and a smooth urethral tube that was easy to negotiate was present in all patients. It was also found that when comparing the effect of Cantwell-Ransley repair on bladder capacity to that of the Young epispadias repair, there was no difference in the eventual bladder capacity (within 24 months after repair: 56 cc versus 54 cc respectively).

DISCUSSION

Many surgical techniques have been described for the reconstruction of the penis and urethra in patients with classic bladder exstrophy or epispadias. Historically, most repairs stem from Cantwell's original description in 1895. Cantwell mobilized the urethra from "Its unnatural attachments", which most likely were the suspensory ligaments. The corporal bodies were separated and urethra transposed ventrally. The corpora were then approximated in the midline. While these early pioneering steps in penile reconstruction met with initial acceptance, devascularization of the urethral closure occurred and led to its abandonment.[1]

Young modified Cantwell's technique by leaving the urethral plate attached to one corpora and transposing this ventrally.[1] Later, Swenson used a buried Denis Browne strip in an attempt to lessen the complication rate.[6] Formerly, the modified Young repair as described by Lepor was the author's choice for construction and coverage of the neourethra.[7] However, our experience with fistulae and with the persistence and recurrence of dorsal chordee has led us to abandon this procedure.

Ransley et al.,[8] introduced the concept of releasing dorsal chordee by incision and anastomosis of the dorsal medial aspect of the corpora over the urethra and ventral meatotomy at the distal end of the glans (IPGAM maneuver) to move the urethral meatus to a more normal position. Proper performance of the IPGAM maneuver is essential if both a normal-appearing glans and good direction of the urinary stream is to result. The dissection of the urethral strip to the midpoint between the corona and meatus provides a ventral position of the urethra in the glans. This approximation of broad, raw surfaces of glandular tissue dorsal to the urethra is clearly the reason why the incidence of fistulas in the area of the corona is practically non-existent as compared to the Young repair.

Fistulas in our group of patients all uniformly appeared at the point of weakness in the base of the penis where the urethra comes up proximally between the corporal bodies. Initially, the authors thought that this would be the most common area of fistulization if paraexstrophy flaps had been previously used. However, fistulas occurred more often in those with an intact urethral plate and in all patients they occurred at the penile base. While the authors agree with Borzi and Thomas that preservation of the urethral plate at the time of exstrophy closure is ideal, we have not found that these paraexstrophy flaps in this selected group of patients, endangered the epispadias repair to any appreciable degree.[9] Nonetheless, great care must be taken to get the dissection directly on the corpora and to keep these previously mobilized flaps thick and well-vascularized.

Another lesson learned with increasing experience is that the Cantwell-Ransley repair effectively corrects intrinsic corporal chordee. While our review of these patients finds that all are straight or deflected downward, many of these patients are still young children. Whether simple rotation of the corpora is adequate or corporal incision and anastomosis is required is really a judgment call at the time of surgery. If the urethral plate is intrinsically short or scarred from prior surgery, then a graft will be required for urethral replacement. Whether patients will require further straightening as they achieve sexual maturity is unknown, but will require prolonged follow up.

Our experience with older children finds that mobilization of the urethral plate and corporal rotation dorsally are inadequate to completely correct the chordee and that incision and anastomosis of the corpora or dermal grafts are required to achieve adequate straightness in these older patients.

With the urethra under the corpora, the urethra attains a more anatomical and straighter course and is supported dorsally by the more normally placed corporal bodies. This has become increasingly important because if intermittent catheterization is needed to achieve continence, an easily catherterizable neourethra might obviate the need for a continent urinary stoma elsewhere. However, reliable intermittent catheterization through a reconstructed bladder neck can still be a problem even with an easily negotiable neourethra.

Another lesson learned in this group was the management of the neurovascular bundle. In those patients in whom corporal rotation was used without corporal incision and anastomosis, the neurovascular bundle was left intact and not dissected from its bed. This variation posed no problem as far as rotation of the corpora medially over the urethra was concerned. However, if incision and anastomosis is needed, then mobilization of the neurovascular bundle is required.

The Cantwell-Ransley procedure can also be used as a salvage procedure. In this group of patients, 3 exstrophy patients and 2 epispadias patients had undergone a prior urethroplasty. All of these patients were able to undergo Cantwell-Ransley repair after testosterone stimulation without difficulty. The most interesting group of patients were those who underwent both reclosure of their bladder exstrophy and combined Cantwell-Ransley repair. All of these patients underwent an anterior innominate osteotomy along with their urethroplasty and reclosure of the bladder exstrophy. The exposure obtained with reclosure of the bladder allows an excellent approach to any remaining suspensory tissue which remains from the initial exstrophy closure. Also, one can get to the proximal limit of the corpora as they come off the pubis so that more urethra can be submerged below the corpora and better correction of the chordee may occur. Then, multiple layers of soft tissue can be interspersed over the proximal urethra and proximal diversion is effected for several weeks which will lower fistula formation.

Another surprising lesson which was learned was the absence of any effect of the epispadias repair technique used, on the eventual bladder capacity. The authors had assumed that there would be increased outlet resistance with the urethra under the corpora and thus, an increased eventual bladder capacity over the Young repair, whereas capacity was equal in the two groups.

This preliminary assessment of urethral reconstruction using the Cantwell-Ransley procedure has shown that in both patients with bladder exstrophy or epispadias, a relatively straight penis with a urethra of adequate caliber is achieved with minimum morbidity. While our fistula rate is a bit higher than our initial communication with this technique,[3] we have included a more complex group of patients, many with prior operative procedures. Yet, the fistula rate of 15%, is still somewhat lower that of the Young repair which creates a 21% fistula rate at our institution.[7] While there is initial data from a few patients, as far as genital function is concerned, a definitive assessment must be deferred until these boys are both sexually mature and sexually active.

In conclusion, the Cantwell-Ransley urethroplasty offers better cosmesis, a straighter penis and urethra and a lower fistulae rate as compared to the Young urethroplasty. As more experience has been gained with this procedure, we have learned to apply it to a wider number of patients in the exstrophy/ epispadias spectrum with more complex anomalies including the exstrophy reclosure patient. Long term follow up is still required for accurate appraisal and to ensure long term success.

REFERENCES

1. Young, H.H.: An operation for the cure of incontinence associated with epispadias. J. Urol., 7:1, 1922.
2. Gearhart, J.P. and Jeffs, R.D.: Complications of exstrophy and epispadias surgery. In: Complications of Urologic Surgery, 2nd ed. Edited by R.B. Smith and R.M. Ehrlich. Philadelphia: W.B. Sounders Co., chapt. 41, p. 569, 1990.

3. Gearhart J.P., Leonard M.P., Burgers J.K. and Jeffs R.D.: The Cantwell-Ransley Technique for repair of epispadias. J. Urol 148:851, 1992.
4. Duckett, J.W.: Use of paraexstrophy skin pedicle grafts for correction of exstrophy and epispadias repair. Birth Defects, 13:175, 1977.
5. Cantwell, F.V.: Operative treatment of epispadias by transplantation of the urethra. Ann. Surg., 22: 689, 1895.
6. Swenson, O.: Pediatric Surgery. New York: Appleton-Century- Crofts, Inc., 1958.
7. Lepor, H., Shapiro, E. and Jeffs, R.D.: Urethral reconstruction in boys with classical bladder exstrophy. J. Urol., 131:512, 1984.
8. Ransley, P.G., Duffy, P.G. and Wollin M.: Bladder exstrophy closure and epispadias repair. In: Operative Surgery- Pediatric Surgery, 4th ed. London: Butterworths, p. 620, 1989.
9. Borzi P.A. and Thomas D.F.M.: Cantwell-Ransley epispadias repair in male epispadias and bladder exstrophy. J. Urol. 151:457, 1994.

22

THE IMPACT OF URODYNAMICS ON SURGERY FOR INCONTINENCE IN CHILDREN WITH EXSTROPHY

Stuart B. Bauer

Children's Hospital
Boston, Massachusetts

The urologic management of the exstrophy patient has always been a daunting experience for clinicians. Following closure, these children often have continued to suffer from urinary incontinence. The decision to enlarge the bladder and/or increase urethral resistance has been a difficult choice for the urologic surgeon, based solely on clinical grounds and radiographic assessment. The ability to accurately predict which procedure singly or in combination would achieve continence has never been very high. In recent years urodynamic testing (UDS) has become an integral part of the preoperative evaluation of the exstrophy patient following closure, especially if incontinence remains an issue. The findings in children with intractable wetting treated at Children's Hospital is presented to provide guidelines for managing this complex anomaly.

Over the past 10 years we have evaluated 58 children with exstrophy (45 simple and 13 cloacal), who underwent closure of the bladder in the newborn period and subsequently had urodynamic studies (either cystometrography [CMG] or urethral pressure profilometry [UPP]) because of continued incontinence. There were 34 females and 24 males. Of the 45 exstrophy patients, 29 were studied before any secondary surgery. 19 of these had CMGs, 12 or whom were normal, 6 were areflexic and only 1 had uninhibited contractions (UC). These 19 also had UPPs, 11 of which were normal (≥50 cm H_2O), while 8 were low (20–30 cm H_2O). After a Young-Dees type bladder neck reconstruction (BNR), most children developed either a hypertonic and/or hyperreflexic bladder while urethral resistance did not rise sufficiently to improve continence. Of the 16 remaining patients with exstrophy who underwent a bladder augmentation prior to UDS, UPPs were low in all but improved to acceptable levels (≥40 cm H_2O) in 60% after BNR. Thus, the detrusor seems to function relatively normally in 2/3 of children following early closure but hypertonicity and hyperreflexia can develop after extensive BNR, necessitating augmentation cystoplasty (AC) to maintain an adequate organ for storage. Consequently, if BNR is contemplated, an AC may need to be performed as well in children with poor detrusor compliance

The Exstrophy–Epispadias Complex, edited by Gearhart and Mathews.
Kluwer Academic / Plenum Publishers, New York, 1999.

in order to achieve a reasonable degree of continence. If AC is performed first, this may alter detrusor muscle orientation to produce a low UPP, which can be improved upon with a later BNR. Thus, BNR may need to be considered simultaneously when AC is undertaken in those children with modest or low levels of bladder outlet resistance.

Of 13 patients with cloacal exstrophy, 7 were evaluated before any surgery other than closure and correction of their meningocele. 3 of the 5 with CMGs were areflexic and 4 of the 5 had low UPPs. Following BNR the bladder remained areflexic and the UPP did not improve in most. Thus, the neurologic injury to the bladder and pelvic floor musculature did not auger well for an effective BNR despite maintaining good detrusor compliance. Of the 6 children with cloacal exstrophy who had an AC prior to UDS, the UPP was low in most and improved in only 3. Thus, BNR might not be of benefit in the cloacal exstrophy child undergoing AC. It seems that alternative surgeries are needed to achieve a reasonable continence mechanism in these children.

23

THE USE OF SUB-MUCOSAL COLLAGEN IN THE SMALL BLADDER ASSOCIATED WITH BLADDER EXSTROPHY

Paolo Caione and Alberto Lais

Bambino Gesù Children's Hospital
Division of Pediatric Urology
Rome, Italy

PATHOPHYSIOLOGY

A small bladder plate is a common finding in the exstrophic complex. However, today it is usual practice to attempt closing at birth each exstrophic bladder encountered, even the very small ones. This principle was first emphazised by the John Hopkins' Pediatric Urologists and later accepted by most worldwide.[1]

After primary closure, the bladder does grow in 30 to 60% of cases, whereas some of these bladders fail to develop adequately. For this reason, in the eighties, Robert Jeffs inverted his sequence of staged functional reconstruction, i.e., suggesting to repair the epispadias during the second or third year of life, before bladder neck surgery, in order to achieve some outlet resistance to let the bladder stretch and grow.[2] This principle has been extended also to the female patient, in whom genitourethroplasty was recommended either at birth during primary closure, or thereafter, during the first 18 months of life.[3]

The combination of bladder closure and reconstruction of the external genitalia, which must be completed early in lifetime, may give nice results in terms of gained functional capacity of the bladder reservoir. This phenomenon sometimes does occur all of a sudden, but usually it is a quite slow process that takes place during the following three to four years. Therefore, it is appropriate for the surgeon to wait and see, closely watching eventual impairment of renal function. In fact, after bladder closure and because of the increased outflow resistance, vesico-ureteral reflux will ultimately occur, leaving those kidneys at risk for subsequent nephropathy.

It is accepted that a functional capacity of 80 to 100 ml is adequate to perform surgery on the bladder neck for urinary continence, and this should be achieved before school age for psychological reasons.[5] We have, therefore, a time period of 4 years maximum to wait, during which time these children must be monitored yearly with cystoscopy and cystogram under general anesthesia, and semiannually with ultrasonography of the upper urinary tract.

The Exstrophy–Epispadias Complex, edited by Gearhart and Mathews.
Kluwer Academic / Plenum Publishers, New York, 1999.

On the contrary, if the bladder fails to develop, other alternatives must be taken into account: bladder augmentation simultaneously or subsequently to the bladder neck reconstruction, bladder substitution, urinary diversion. These alternatives are costly, both under the clinical and social standpoint, and should be considered as the last resort.

The failure of bladder growth may be due to two main reasons:

1. Histology changes of the bladder wall

We have recently looked at the distribution of smooth muscle and collagen in the exstrophic bladder at different stages of functional reconstruction. From our data it appears that the percentage of smooth muscle increases substantially, following primary closure, urethroplasty and bladder neck reconstruction. However, smooth muscle does not reach a distribution equal to that of normal controls of similar age.[6] These data demonstrate that the bladder that has never acquired function in utero maintains some deficient capability to develop totally after refunctionalization.[7] Studies on the subtypes of collagen in the exstrophic bladders have been carried out by the John Hopkins group recently.[8] In this study types I and type III collagen have been found increased in the newborn with exstrophy. Some concern in this study appears from the additional finding of increased type II collagen, which was not known to be present in the bladder.[8]

The consequences of these histologic changes, or fibrosis, are ultimately a small capacity, low compliant bladder.

2. Low outflow resistance

In bladder exstrophy surgery, it is not infrequent that the outflow remains wide open, despite primary closure and urethral reconstruction. In such cases the *leak pressure point* (LPP) is proximal to zero, and the bladder wall is not stimulated to function.

A new concept, *"the never functioning bladder"* has thus developed.[7,9] That is, the bladder that has never acquired its function prenatally or early postnatally, and therefore has modified its histologic pattern as previously described.[9] Such bladders remain small, and we have shown that surgery for continence and ureteral reimplantation may be procedures at risk in these patients.[7]

In the aforementioned situation, which may represent up to 40–50% of the exstrophic bladders, the aim is to increase the outlet resistances, either by open surgery (re-do Young Dees Leadbetter) or by endoscopic techniques. At least, the bladder capacity which can be achieved after primary closure seems to be the most important factor related to successful continence surgery in the exstrophy patient.[10]

INDICATIONS

In the exstrophy-epispadias complex, periurethral submucosal collagen injection may be indicated either before reconstruction of the bladder neck, after urethroplasty, or conversely after the Young-Dees operation, should the latter proved insufficient.

The most appropriate timing for periurethral collagen injection in our opinion, is *before* completion of the functional reconstruction. The rationale for its use in this case is to increase the outlet resistance and let the bladder stretch and gain volume. The procedure can be repeated over time should it be necessary, and it does not interfere with further surgery on the bladder neck.

The second main indication for submucosal periurethral collagen injection in this group of patients is the type III stress incontinence, which may be present as consequence of intrinsic deficiency of cervico-urethral resistance after partial failure of previous Bladder Neck reconstruction. Collagen injection is an alternative to a re-do operation and may be repeated up to two or three times.[17]

Periurethral-submucosal injection of collagen should not be regarded as a complete substitution for bladder neck surgery in exstrophy patients. It may be the only form of treatment for continence in minor degree of epispadias, in which there is a solid backing in the cervical area.

TECHNIQUE

The substance most frequently used is glutaraldehyde cross linked bovine collagen (Zyplast®, Contigen®, Mediplast®), although other substances are used currently and have been used in the past. Among these are polytef paste (Teflon®), silicon, autologous fat, and more recently dextranomer in jaluronic acid (Deflux®). Deflux® has given us good results in the endoscopic treatment of vesico-ureteral reflux[12] and is promising for its use for continence too.

When collagen is used, a skin test is administered 4 weeks prior of the implant to rule out possible hypersensitivity to the substance. Specific informed consent is obtained from the parents or tutors. Parents should be informed about different options currently offered by open surgery, and possible immediate or long term adverse reactions. Also, information should be given of the possibility of failure and/or need to repeat the treatment.[13]

The patient is prepared as for any diagnostic cystoscopy. We use a Storzl cystourethroscope 11, 12 or 14 Fr., with zero degree lens. A 30 degree lens can be sometimes useful. A specially designed needle is used to deliver the implant material submucosally.

The operative technique may be *submucosal trans-urethral* or *periurethral*.[14,15] In the first method, the needle is advanced through the operative channel of the endoscope and across the mucosa of the bladder neck or posterior urethra. The injections are carried out on three different sites, usually 3, 9 and 12 o'clock. It is important to keep clear of the 6 o'clock position in the male patient, not to injury the verumontanum and developing prostatic tissue. At each site 0.75 to 3 cc of substance are injected until visual occlusion of the lumen is achieved. An average of 7.5 cc is usually required in each patient. This method is preferred in males and after previous bladder neck surgery.

The *peri-urethral* technique is used rarely and almost invariably in females after primary closure of the exstrophic bladder. The needle is inserted at the side of the urethral meatus and advanced through the external sphincter up to the mucosa where it can be seen buldging through the endoscope.

Antimicrobic prophylaxis is maintained for 2 to 3 days following endoscopy, or longer if reflux is present. Anticholinergic drugs (Ditropan®) are useful for 2–3 months after the procedure to avoid latent bladder hypercontractility and/or hypertonicity, which may become evident after the increase of the outflow resistance.

RESULTS

In judging the results one must keep in mind the two different goals of the endoscopic collagen injection on the bladder neck of the exstrophic bladder.

As this procedure does not represent an alternative to bladder neck surgery *per se*, results on *continence* may be given only for those cases injected after complete functional

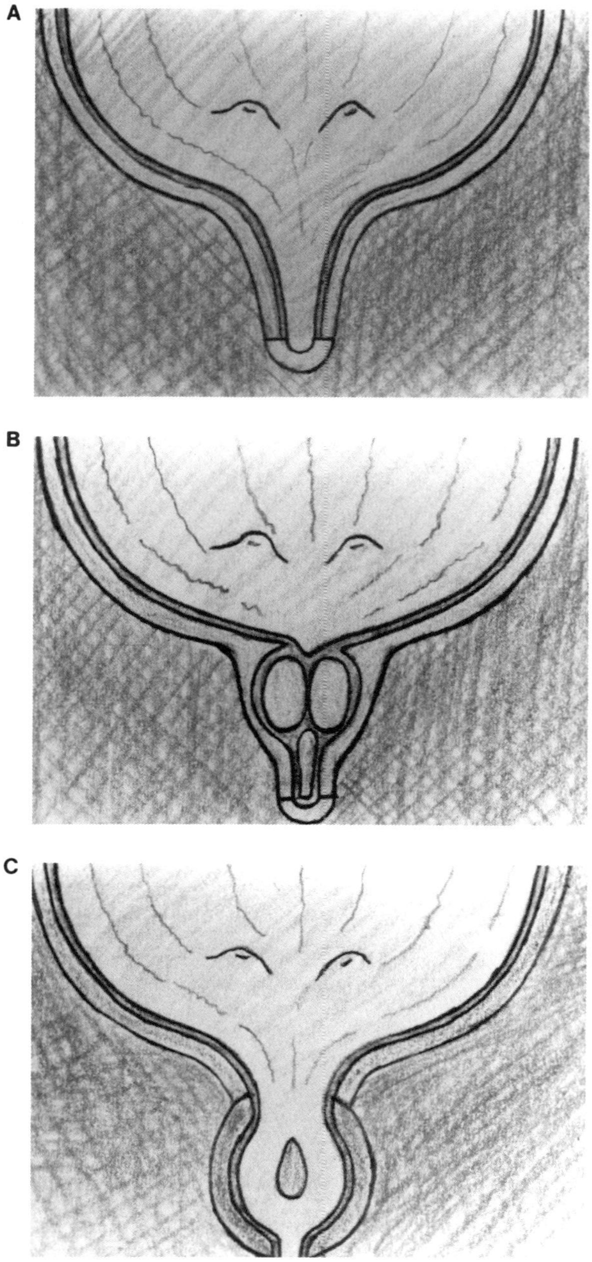

Figure 1. Periurethral submucosal injection of bulking substance for increasing poor outlet resistance. A (top): Female patient. B (middle): injection on bladder neck. C (bottom): male patient.

reconstruction of the bladder. We treated 17 such children. Six of these resulted in 3 or more hours of dryness, 5 in 1 to 2 hours, and 6 were failures (less than 1 hour of dryness).

When collagen injection is performed before bladder neck closure, the aim is to increase the outlet resistance and let the bladder stretch. In such cases results are not judged by achievement of continence, but rather as increase of bladder capacity. We have treated 9 such cases (age range 1 to 4 years) at our Institution with a mean increase in capacity of 75% in 6 cases and little or nil gain in 3 children, after one to three injections.[16]

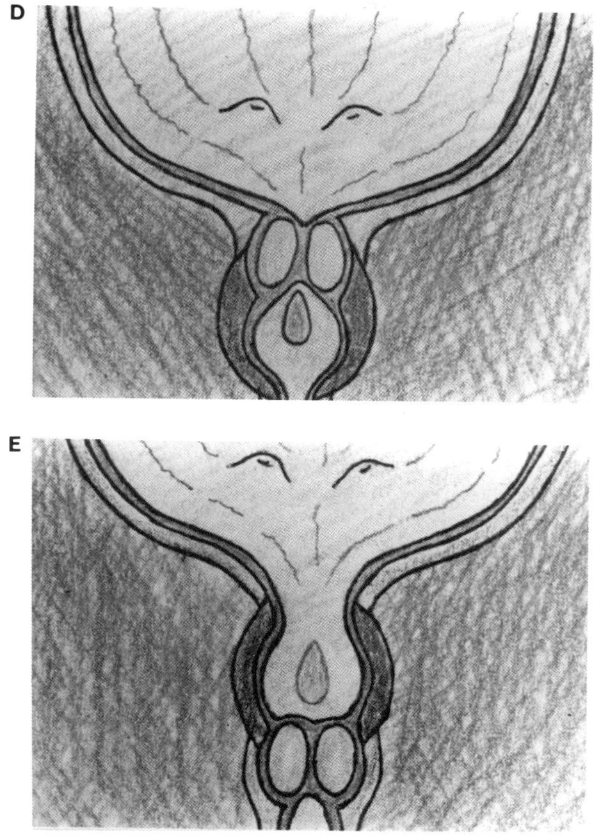

Figure 1 (*cont.*). D (top): injection on bladder neck. E (bottom): injection on infrasphincteric urethra.

COMPLICATIONS

Basically, three orders of complications may occur, being responsible of partial or complete failure. These are a technical problem, i.e., the *spillage* of the substance in the lumen of the urethra during the endoscopic injection. The spillage may occur as consequence of the scarring of the submucosa after previous surgery on bladder neck. On the long run, the reabsorption of the watery component and the biodegradability of the substance ultimately result in a *volume reduction of the implant.* The volume reduction is more evident for the 35% collagen components, but a minor reabsorption in seen also when we use the 65% collagen composition matherial or the dextranomer with jaluronic acid. A third cause of longterm failure is the *secondary displacement* of the implanted material from the original site of impiant. In this event, the dislocation occurs from the implant site along the lines of minor resistance,[13] ultimately resulting in an incomplete occlusion of the lumen, being the implants on different planes of the posterior urethra or bladder neck.

CONCLUSIONS

What are we to expect from this mininvasive technique? Of course, we should not expect the global cure of both the incontinence and of the small capacity which is usually seen in the exstrophy-epispadias complex. However, it can be very helpful in achieving both

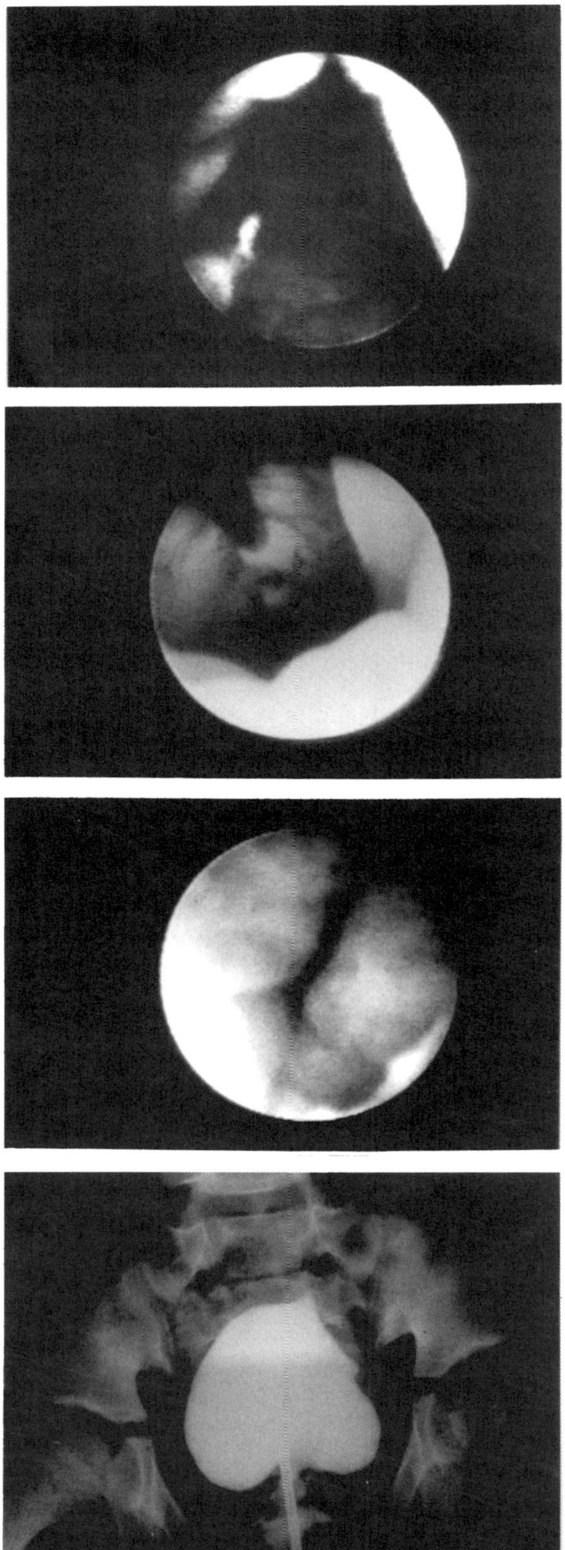

Figure 2. Endoscopic trans-uretral injection for continence. A (top): wide bladder neck. B (first from top): injection at 12 o'clock. C (second from top): good visual occlusion of theurethral lumen has been achieved with 4 sites injections. D (bottom): radiographic picture (cystogram) during the procedure.

these goals by adding little and repeated progress at a time. The exstrophic bladder is ultimately a reservoir that needs to exercise itself through progressive refunctionalization.[16,17]

In this perspective, the endoscopic procedure on the bladder neck or urethra allows the increase in outlet resistance that may prove adequate for bladder growth and maturation, obviating the need for augmentation at a future date in a number of cases.

Mininvasive techniques are presently evolving in many surgical fields, gaining their space over conventional open procedures. This is also the case with endoscopic treatment of urinary incontinence. In children, one must adapt the technique to the peculiarity of the anatomical situation. Also the basic principle is that of tissue substitution, but as yet the ideal bulk enhancing substance is to be defined.

REFERENCES

1. Jeffs R.D. : Exstrophy And Cloacal Exstrophy. Urol Clin North Am. 5:127, 1978.
2. Oesterling J.E., Jeffs R.D.: The Importance of a Successful Initial Bladder Closure in the Surgical Management of Classical Bladder Exstrophy: Analysis of 144 Patients Treated at The John Hopkins Hospital between 1975 and 1985. J. Urol. 137:258, 1987.
3. Gearhart, J.P. Jeffs R.D.: Bladder Exstrophy: Increase in Capacity Following Epispadias Repair. J. Urol. 142:525, 1989.
4. Caione P., Capozza N., Lais A., Creti' G., De Gennaro M.: Female Genitourethroplasty and Submucosal Periurethral Collagen Injection as Ajunctive Procedures for Continence in the Exstrophy Epispadias Complex. Br. J. Urol. 71: 350–353, 1993
5. Duckett, Caldamone A.: Bladder and Urachus In: Kelalis P.P., King R.L., Belman A.B. (eds): "Clinical Pediatric Urology. Philadelphia, W B Sanders Co., 1985, Pp 726–751.
6. Lais A., Paolocci N., Ferro F., Bosman C., Boldrini R., Caione P.: Morphometric Analysis of Smooth Muscle in the Exstrophy-Epispadias Complex. J. Urol. 156: 819–821, 1996.
7. Lais A., Caione P., Rossi L., De Gennaro M., Capozza N.: The Never Functioning Bladder; Pitfalls of Vesico-Ureteral Reimplants. Proceedings of the European Society of Paediatric Urology (espu) 5th Annual Meeting, Goteborg (Sweden) 28–30 April 1994 (abstract).
8. Lee B.R., Perlman E., Partin A.W., Gearhart J.P.: An Analysis of Collagen Subtypes in the Newborn Exstrophy Bladder. Presented at the American Academy of Pediatrics. Section on Urology, Meeting, San Francisco, CA, October 14–16, 1995 (abstr. 124, Pag. 185)
9. Lais A., Bosman C., Rossi L., Boldini R., Caione P.: Definition of the "Never Functioning Bladder". Proceedings of The 12th Congress of the Asian Association of Pediatric Surgeons; Hong Kong 6–9 November 1994, Abstr. 19.4.
10. Caldamone A.A.: Anomalies of the Bladder and Cloaca. In Gillenwater J.Y., Grayhack J.T, Howards S.S., Duckett J.W. (eds): "Adult and Pediatric Urology", Year Book Medical Publishers Chicago, 1987, Pp 1809–1835.
11. Mc Guire E.J., Appel R.A.: Transurethral Collagen Injection for Urinary Incontinence. Urology 43:413–415, 1994.
12. Capozza N., Nappo S., Ferro F., Caione P.: Trattamento Endoscopico Del Reflusso Vescico-ureterale: Due Diversi Materiali Biodegradabili A Confronto. 32th Congress Italian Society of Pediatric Surgery. Turin 25–28 Sept. 1996 (abstr)
13. Capozza N., Caione P., De Gennaro M., Nappo S., Patricolo M.: Endoscopic Treatment of Vesico-Ureteric Reflux and Urinary Incontinence: Technical Problems in the Pediatrics Patient. Br. J. Urol. 75:538–542, 1995.
14. Caione P., Lais A., De Gennaro M., Capozza N.: Glutaraldehyde Cross-linked Bovine Collagen in Exstrophy-Epispadias Complex. J. Urol. 71: 350–352, 1993.
15. Caione P.: Endoscopic Approaches to the Treatment of Incontinence In "Smith's Textbook of Endourology", Arthur D. Smith Ed., Qmp, Quality Medical Publishing. Inc Editor, St Louis MO, 1996 Chapter 85 Pagg. 1338–1351.
16. Caione P., Lais A., Capozza N., Matarazzo E., Ferro F., Rossi L.: Endoscopic Treatment of Structural Sphincteric Incompetence for Continence and Bladder Volume Enhancement: A Lesson from 7 Years Experience 8th Espu Annual Meeting, Rome April 3–5, 1997.
17. Bomalaski M.D., Bloom D.A., Mc Guires E.J., Panel A.: Glutaraldheyde Cross-linked Collagen in the Treatment of Urinary Incontinence in Children. J. Urol. 155:699–702, 1996.

24

REVIEW OF LONG-TERM RESULTS OF STAGED FUNCTIONAL RECONSTRUCTION AND BLADDER NECK PLASTY IN BLADDER EXSTROPHY PATIENTS FOLLOWED FROM BIRTH AT THE JOHNS HOPKINS HOSPITAL

Yegappan Lakshmanan, John P. Gearhart, and Robert D. Jeffs*

The Johns Hopkins Hospital
James Buchanan Brady Urological Institute
Division of Pediatric Urology
Baltimore, Maryland 21287

HISTORICAL PERSPECTIVE

Bladder neck plasty, which is the planned third stage in staged reconstruction of bladder exstrophy grew out of the earlier contributions to incontinence surgery by Hugh Hampton Young, John Dees, Guy Leadbetter and Victor Marshall. Early attempts at reconstruction of classical bladder exstrophy were often dismal failures leading to the statement in the reigning pediatric urology text by Campbell[1] in 1951 to say that all exstrophies should be diverted to the skin or the bowel as their best means of management.

Significant contributions to reconstruction by Sir David Innes Williams, Herbert Johnston, John Lattimer and Tague Chisholm led to a reawakening of attempts to reconstruct the exstrophy. Many innovations have followed their early lead. In most instances, the early attempts to reconstruct the exstrophy were done as a single-stage approach, with closure, epispadias repair and reconstruction of bladder neck area for continence, all in one procedure. Schultz[2] and Shusterman introduced the concept of performing posterior iliac osteotomy to facilitate the apposition of the pubis at the initial closure.

---

* Contact author: Robert D. Jeffs, M.D., Professor of Pediatric Urology, Marburg 1, James Buchanan Brady Urological Institute, The Johns Hopkins Hospital, 600 N. Wolfe Street, Baltimore, MD 21287-2101

The Exstrophy–Epispadias Complex, edited by Gearhart and Mathews.
Kluwer Academic / Plenum Publishers, New York, 1999.

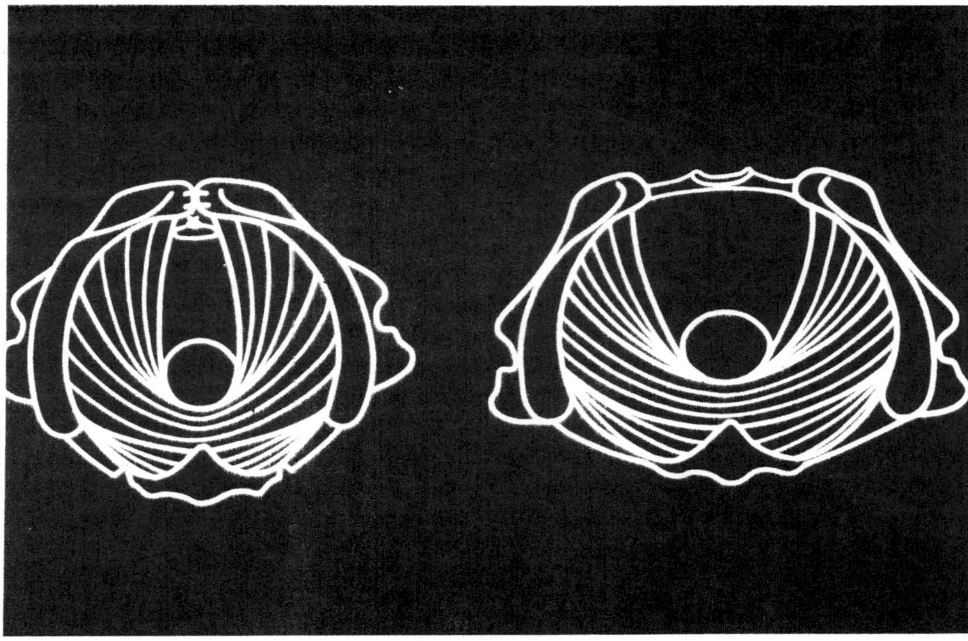

Figure 1. Pelvic ring before and after exstrophy closure, illustrating the position of the posterior urethra and pelvic musculature.

These early attempts at bladder reconstruction in a single stage did, however, indicate that the bladder could grow, that some patients could become continent and also showed some of the shortcomings, namely, the problem of vesicoureteral reflux and upper tract damage, the frequent occurrence of dehiscence after early closure and the marked deterioration of the bladder mucosa which could occur when left exposed prior to closure for a period of months or years.

In addressing the problem of dehiscence of the bony and soft tissue closure in front of the bladder, it became evident that external spica plaster did not adequately prevent the distracting muscle forces that these vigorous young babies could exhert in the closure area. The "mummy-wrap" more recently introduced also fails to protect the anterior closure in the critical early weeks of healing.[3]

The posterior iliac osteotomy was introduced with the idea of allowing the bony pelvis to come into apposition at the symphysis, thus providing protection for the closed bladder and urethra underneath and relieving the soft tissue tension in the overlying fascial and skin closure. In addition to the benefit in the closure phase, it also theoretically would suggest that bringing the pelvic floor muscles together might have a bearing on the eventual continence (Figures 1 and 2). This is given some support in review of older patients, who not only have a period of dryness from their bladder neck closure, but also are able to stop and start their stream, presumably with the help of these pelvic floor muscles.[4]

The orthopedic input from surgeons such as Salter and Hall suggested that postoperative immobilization be carried out in gallows traction, or in slightly older children, in modified Bryant's traction, to reduce the distracting forces at the anterior closure and did appear to reduce the incidence of dehiscence after the initial closure (Figure 3).

With closure of the exstrophy bladder, resistance to urinary flow is introduced and it soon became evident that if this was too tight, there would be back pressure causing hy-

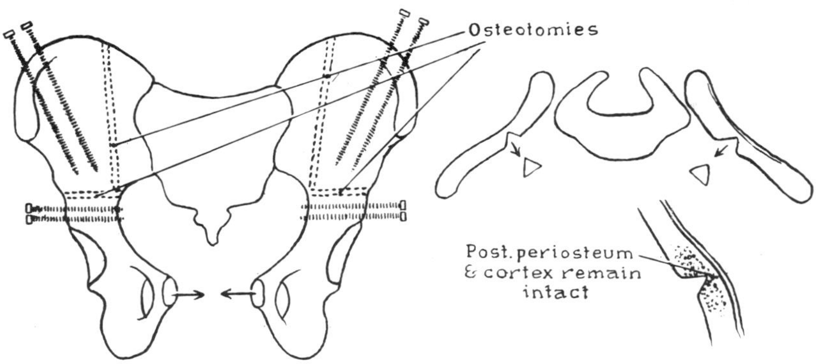

Figure 2. Osteotomy and position of pins with preservation of posterior periosteum and cortex.

dronephrosis and even in those without excessive pressure, the high incidence of reflux, which is nearly 100% in the classical bladder exstrophy, became a problem in managing urinary tract infection. The contributions of Politano and Wyland Leadbetter and many others to the management of the reflux problem was brought to bear on these patients with bladder exstrophy; thus, in the process of reconstruction, reimplantation of the ureter was carried out to correct reflux.

In early instances of reconstruction, the problems of anesthesia in the very immature child were apparent and consequently, there was a delay in closure, waiting for the child to

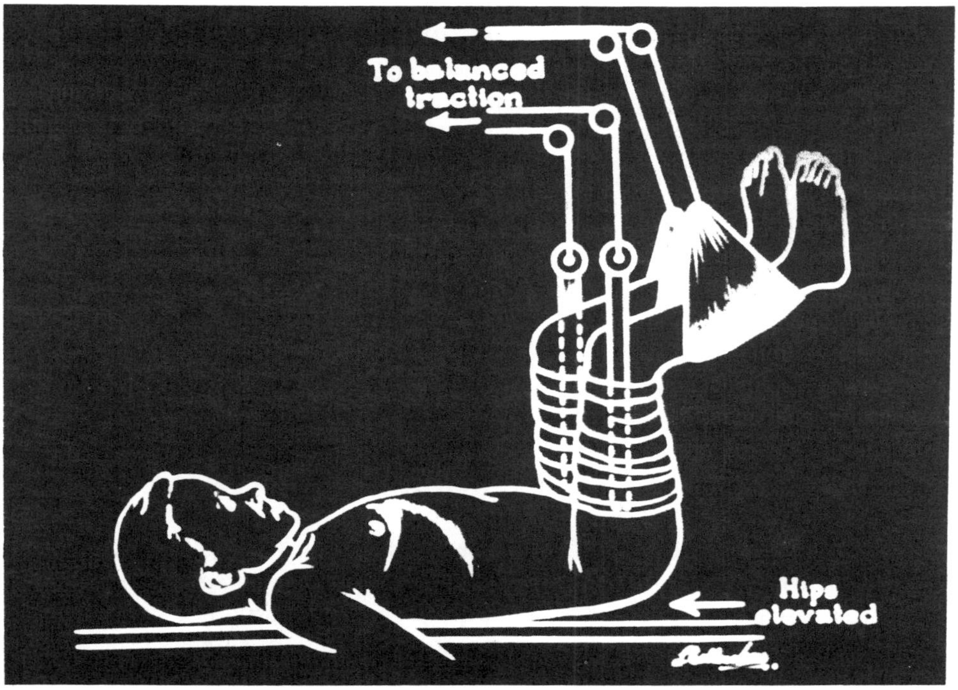

Figure 3. Modified Bryant's traction typically used for immobilization after bladder exstrophy closure without osteotomies. Note the elevation of the hips.

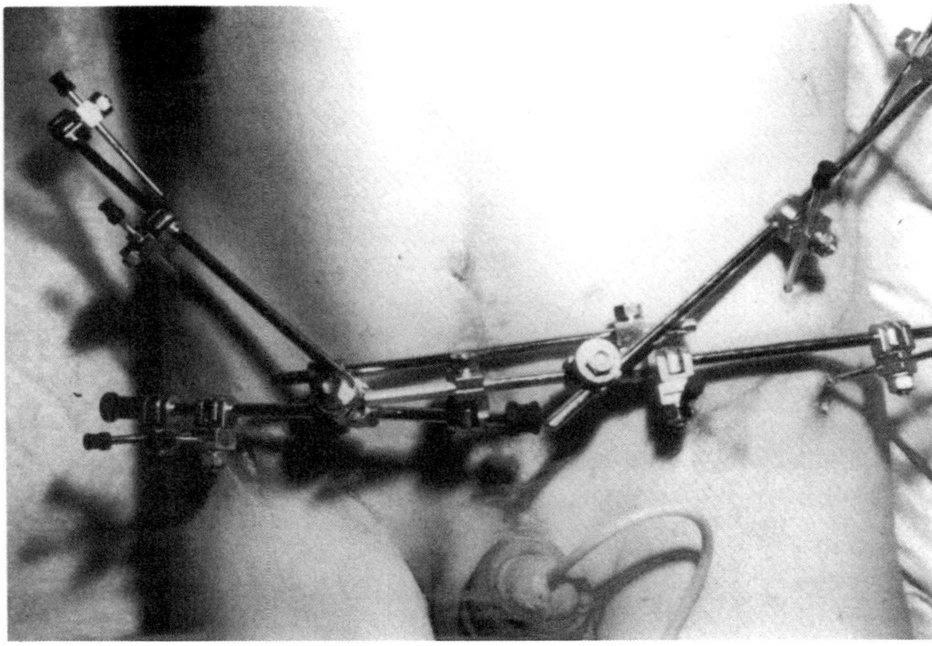

Figure 4. External fixation device used after osteotomies, along with Buck's traction. Shown here is a 2-year-old who had osteotomies at the time of reclosure.

become somewhat more robust and able to stand the rigors of osteotomy and lengthy anesthesia in the process of their closure. With the improvement in management of the newborn and in methods of anesthesia, Julian Ansell[5] suggested that these children be closed at birth before changes in the bladder lining could occur and with the hope that one would have a better chance of the bladder growing and attaining a near normal function.

Coincident with the improvements in the approach to the bladder, similar improvements were made in the epispadias repair and in attempting to give adequate penile length. The present approach to epispadias repair will be described in other chapters in this book.

Anterior innominate osteotomy, first suggested to us by Dr. Doctor, has been further developed by Dr Sponseller, along with the external fixator to provide immobilization (Figure 4). This has become the method of choice in closing the older child or the child with unduly wide diastasis at birth. Dr Sponseller's modification of the anterior osteotomy and the scientific measurements which he has made of the exstrophy pelvic abnormality and its subsequent growth are detailed in other chapters of this book.

EVOLUTION OF STAGED EXSTROPHY RECONSTRUCTION

An experience which occurred at the American Urological Association in 1972 gave impetus to the concept of staged reconstruction. At that meeting, John Lattimer presented his review of patients with reconstructed bladder exstrophy, mostly in a single stage indicating a continence rate of only 4%.[6] A paper following his, by Ormond Culp[7] indicated that the continence rate in patients with complete epispadias after bladder neck plasty was as high as 50%. It seemed logical that if one could close classical bladder exstrophy and allow the bladder to heal, to grow and to increase in volume slowly, while preventing the ravages of infection and back pressure through low pressure drainage of the incontinent outlet, one might then be able to turn the patient with classical exstrophy into one with the features of complete epispadias and later, to perform a bladder neck plasty and achieve a much higher degree of continence, such as Ormond Culp had, in his complete epispadias series.

Out of this thinking grew the plan of staged reconstruction which has been used in the management of our patients, reviewed later in this chapter. Staged reconstruction consisted of a closure at birth, with or without osteotomy, depending on the degree of diastasis and the malleability of the newborn pelvis. Second stage was to repair the epispadias sometime between one and two years of age, thus producing some additional resistance hoping to promote cycling of the bladder and a further increase in its growth and volume.[8] The third stage was to perform the bladder neck reconstruction (BNR) approximately at age four years depending on the bladder's having reached a capacity of at least 60 cc. before that operation would be undertaken.

It also became obvious that it was necessary in the postoperative period, and between each stage, to have intense and continuous follow up to prevent the problems of infection, stone formation, reflux nephropathy, hydronephrosis and retention. Later review recognized the difficulties of fostering the child's ability to recognize his full bladder, to promote a detrusor action and to build on a dry interval, which would become socially useful in its extent and control.

Experience showed that small bladders did grow and could approach the volumes expected for their age. It was also obvious that a dry interval did not occur overnight and that time was required, first for volume to develop in the initially closed bladder and secondly, that the dry interval and consequently the retained volume also took time to develop; one could not expect immediate continence as a general rule after bladder neck plasty.

The third stage, the reconstruction of the bladder neck is undertaken when the bladder has attained a reasonable size for age and a figure of 60 cc. of volume under anesthetic at cystoscopy in a four to five year old child was determined to be the minimum requirement. The essence of bladder neck reconstruction is to re-expose the bladder exstrophy and recognizing its short, broad base one makes a transverse incision which is later closed in a vertical fashion to attempt to narrow this broad base to some extent (Figure 5). With the bladder open, reimplantation of the ureters is carried out, usually in a cephalo-trigonal manner which allows the new, intravesical tunnel of the reimplanted ureter to be sufficiently distant from the bladder neck plasty that the closure of the bladder neck is not interfered with (Figure 6). The mucosal strip starting distally at the veru montanum is selected to be the lining of the new posterior urethra and bladder neck. This mucosal strip includes the underlying trigonal muscle giving the mucosa additional strength and is usually approximately 10 to 12 mm. wide and 3 to 4 cm. in length from the veru montanum proximally (Figure 7). The muscle lateral to the bladder neck is denuded and the denuded distal bladder base is scored by judicious incisions to further narrow the bladder base and provide the overlapping muscle closure which will reinforce the closed mucosal tube (Figure 8). When the bladder neck plasty is completed and the bladder is closed with a suprapubic drain, a pressure test measuring the outlet resistance should be made. If there is still leakage at low pressure, then the outlet should be revised to provide resistance sufficient to withstand a pressure of greater than 50 to 60 cm. of water pressure. The outlet resistance can always be reduced by dilatation subsequently but a tightening procedure is much more complicated and cannot be easily accomplished without reoperation. Drainage of the urinary tract is accomplished by ureteral stents, suprapubic catheter and avoidance of a drain per urethra, which is apt to put pressure on the posterior urethral closure and

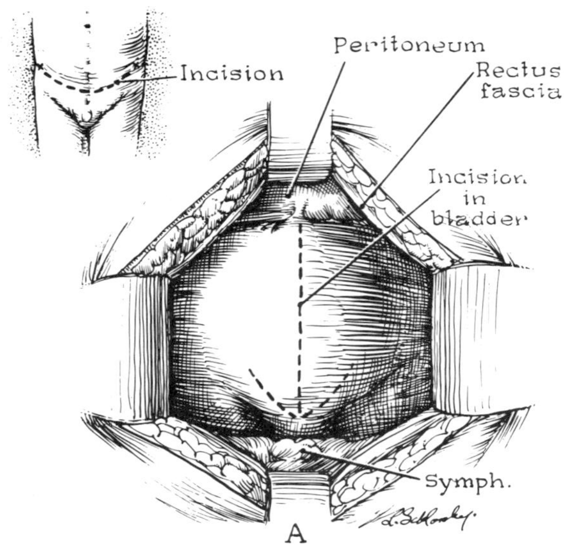

Figure 5. Young-Dees-Leadbetter bladder neck reconstruction: Position of abdominal and 'inverted-T' bladder incisions for optimal access.

consequently, break down some of the resistance that may have been created by the reconstruction at the bladder neck. The final step in bladder neck reconstruction is depicted in Figure 9, showing the suprapubic drainage, a simple manometer measuring outlet resistance and also depicting the suspension of the newly constructed bladder neck to the symphysis or anterior rectus fascia in the manner of Marshall-Marchetti, creating an angulation between the posterior urethra and the bladder floor.

The results to be presented, of the staged management, represent patients presenting at one center to either of two senior surgeons for their first treatment after birth and re-

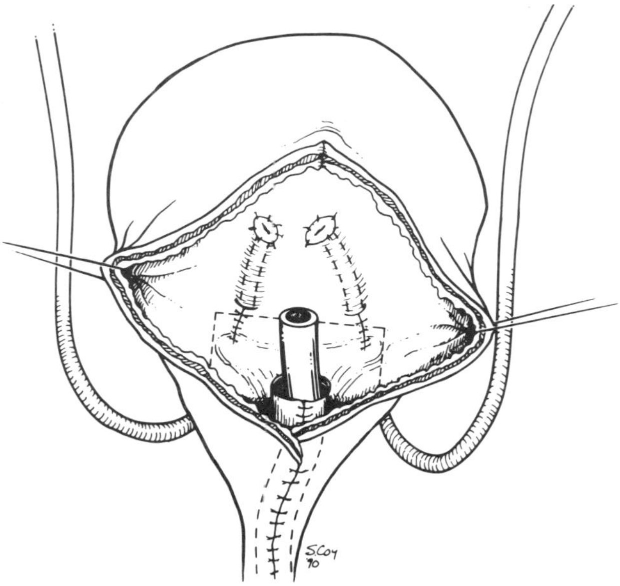

Figure 6. Cephalotrigonal reimplantation of the ureters before incorporation of trigone into Young-Dees-Leadbetter bladder neck reconstruction. Outlined area depicts additional tissue available for tubularization with the ureters directed superiorly.

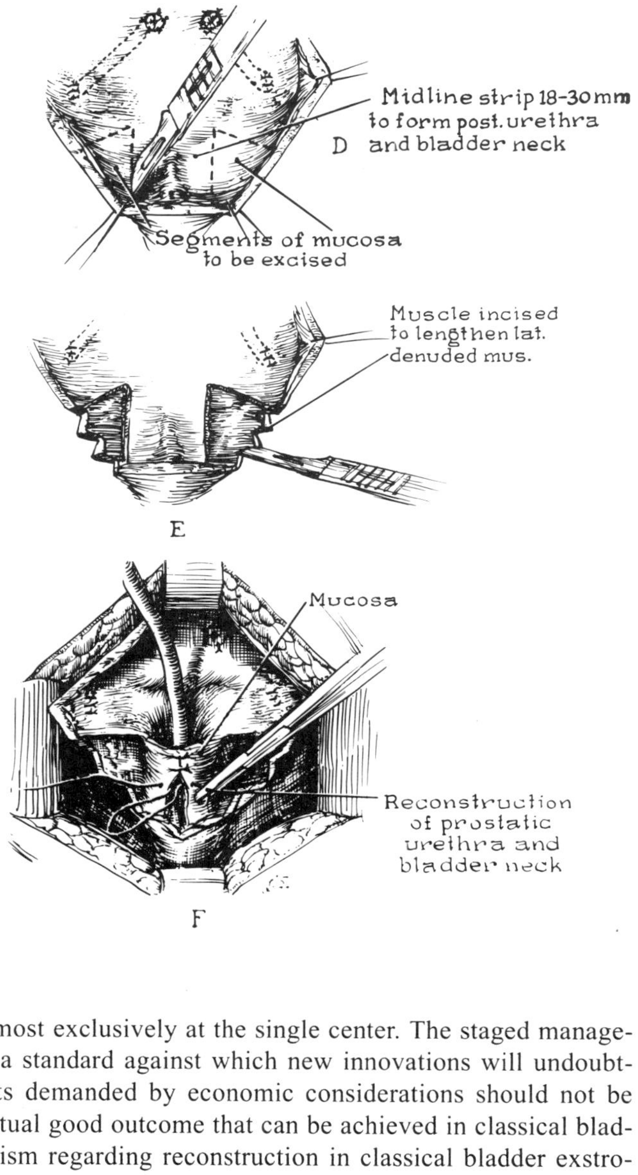

Figure 7. Reconstruction of prostatic urethra and bladder neck.

ceiving their subsequent care almost exclusively at the single center. The staged management of these patients provides a standard against which new innovations will undoubtedly be compared. The shortcuts demanded by economic considerations should not be allowed to detract from the eventual good outcome that can be achieved in classical bladder exstrophy. The early scepticism regarding reconstruction in classical bladder exstrophy surely can be dispelled by the excellent results obtained by so many of the patients in this series and hopefully, this will provide an incentive to improve still further on these results with carefully considered improvements in management and technique.

MATERIALS AND METHODS

This review consists of 88 patients who presented with classical bladder exstrophy, without prior treatment, to the two senior authors between 1975 and 1996. There were 65 boys and 23 girls, with the age at closure ranging from 0 to 5.9 years with a mean of 0.28 years (Table 1). The present age at follow up ranges from 0.83 years to 22.58 years (mean—11.09 years). Two of the 88 patients were not closed, one because the physician

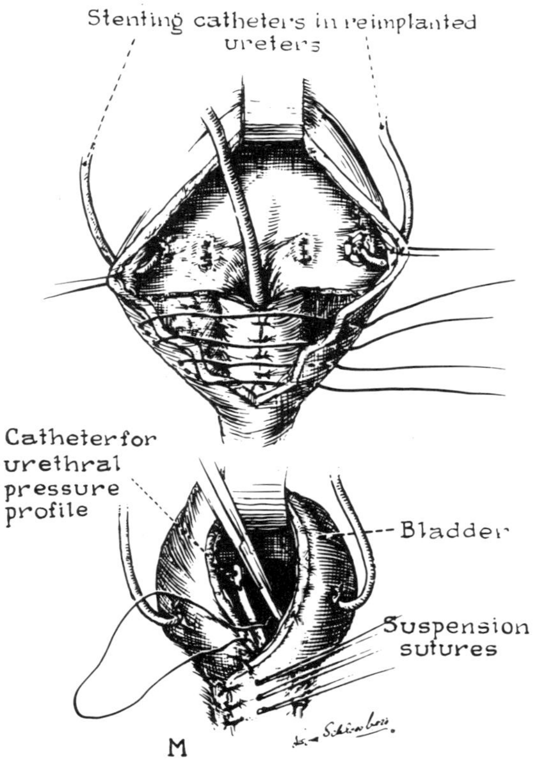

Figure 8. Position of ureteral catheters and Marshall-Marchetti-Krantz type of suspension sutures.

father elected to have a colon conduit and the other from a gypsy family failed to keep many bookings for initial surgery or was unwell at presentation.

Fifty-nine of 64 males have had epispadias repair which was usually performed in advance of BNR to encourage growth in bladder volume by increasing resistance. The average age at epispadias repair was 2.77 years (range 0.92 to 6.5 years). Four patients await epispadias repair and 1 is lost to follow up (Table 3).

Fifty-eight (44 males and 14 females) of the 86 patients, who had their exstrophy closed, developed a bladder capacity of 60 cc or better and were suitable for a subsequent continence procedure. BNR was carried out at an average age of 4.28 years (range of 1.75 to 8.33 years). Of these, it is too early to evaluate continence in 4 patients, who underwent the procedure within the last 6 months. Twenty-eight patients have not had BNR for continence. In this group, 3 patients became continent after initial closure, 2 of whom subsequently required bilateral reimplantation to correct reflux to prevent ascending infection. The third patient is stable without reflux. Twenty-two patients are young and are awaiting improved awareness and increased bladder volume before undergoing BNR. Three patients, with bladders considered unsuitable for functional reconstruction, have undergone augmentation prior to BNR (Table 4).

Postoperative complications were noted from the hospital records. Continence rates are those reported by parents at the time of regular follow-up over the years. For the purpose of this review patients were considered socially continent if they were dry during the daytime for 3 hours or more without stress incontinence, with or without nocturnal continence.

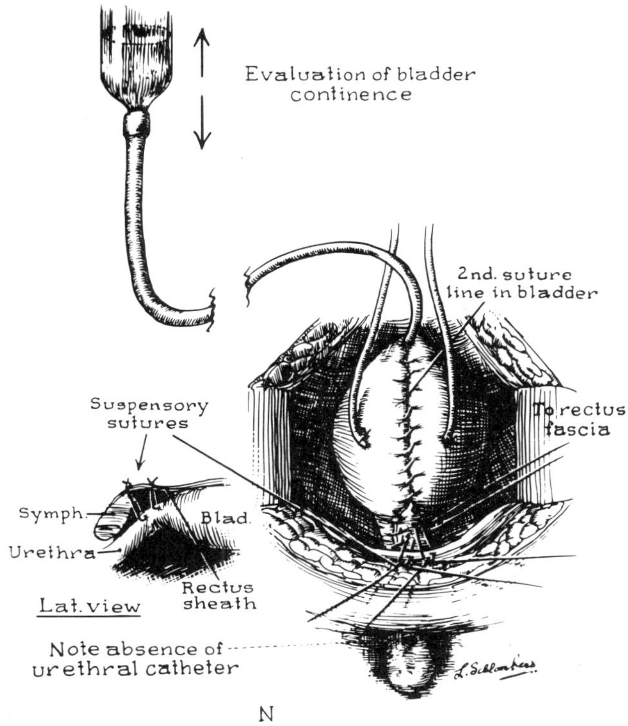

Figure 9. Bladder closure and suspension sutures. Note suprapubic urinary diversion and avoidance of urethral stents or catheters.

RESULTS

Primary Bladder Closure and Osteotomy

Successful closure occurred in 83 of the 86 patients. Three patients had partial dehiscence despite diligent efforts to reduce distracting forces, violent respiratory movement, bladder pain and spasm (Table 2). Two of these were originally closed without osteotomy while one had osteotomy. This latter patient developed respiratory syncytial virus with a severe cough immediately postoperatively leading to partial dehiscence. All of these patients subsequently underwent reclosure with osteotomy.

Osteotomy was used in all patients presenting after 72 hours of age. Osteotomy was performed in 57 patients while 29 were closed without. Five of the 29 subsequently had osteotomy for reclosure or wide separation of the bony symphysis. Posterior iliac osteo-

Table 1. Demographics

Total number of patients	88
Males	65
Females	23
	Age in years (range)
Primary closure	0.28 (0 to 5.92)
Epispadias repair	2.77 (0.92 to 6.5)
Bladder neck reconstruction	4.28 (1.75 to 8.33)
Present age	11.09 (0.83 to 22.58)

Table 2. Primary closure and osteotomy

Total Closures	86
With primary osteotomy	57
Without primary osteotomy	29
Requiring secondary osteotomy	5/29
Total osteotomies	62
Posterior	33
Anterior	26
Pubic ramotomy	3
Complications	
Dehiscence (partial or complete)	3
With osteotomy	1
Without osteotomy	2
Post-op hydronephrosis	2

tomy was used in 33 of the total of 62 patients while anterior osteotomy or a variation of it was used 26 times and anterior ramotomy in 3. Failure to develop a firm fibrous union at the symphysis after initial closure was considered a reason for osteotomy at a later procedure. Only one of the patients closed primarily with osteotomy, mentioned above with respiratory syncytial virus, had a second osteotomy. However, among the 28 patients closed initially without osteotomy, 5 (including the 2 with partial dehiscence) were judged to have an inadequate inter-symphyseal attachment with persisting wide diastasis, requiring osteotomy at a later procedure: one for reclosure, 3 at the time of epispadias repair and 1 during BNR (Table 2).

Paraexstrophy flaps were used in lengthening the urethral plate of the boys who underwent primary closure (41 of 63). The urethral plate was not divided in any of the girls.

Hydronephrosis followed initial closure in 2 patients and was treated by removal of a migrating inter-symphyseal stitch and bilateral reimplantation in one and by cutaneous diversion in the other.

Epispadias Repair

Previously the Young type of epispadias repair was used (27 of 59); due to the high incidence of fistulas noted, this has been replaced in our practice by the Cantwell-Ransley[10] procedure or a modification thereof (32 of 59). A total of 21 patients were noted to

Table 3. Epispadias repair

Closed exstrophy males	64
Epispadias repair completed	59
Awaiting repair	4
Lost to followup	1
Type of repair	
Young	27
Cantwell-Ransley	32
Postoperative fistulas	21
Fistulectomy	14
Spontaneous closure	5
Persisting	2

Table 4. Bladder neck reconstruction

Closed exstrophies	86
BNR completed	58
Continent before BNR	3
Augmented before BNR	3
Awaiting BNR	22

have postoperative fistulas, 14 requiring operative correction. Spontaneous closure occurred in 5, while 2 have persistent fistulas at this time (Table 3).

Bladder Neck Reconstruction

Fifty-eight patients have undergone the modified Young-Dees-Leadbetter type of BNR with the intention of making them continent excluding 3 who became dry without a BNR, 3 who were augmented before achieving a satisfactory bladder volume and 22 patients who are awaiting suitable age and bladder volume before BNR. Four patients have had BNR within the last 6 months and hence, are not evaluable for continence yet (Table 4).

Collagen injection was used to increase resistance in 8 patients; to increase the bladder volume before BNR in 2 and to improve continence after BNR in 6. Each of the 6 became socially continent but collagen to increase bladder volume before BNR was less successful.

Seven patients were treated by augmentation or diversion. In 4 patients the bladder was not felt to be large enough or suitable for a continence procedure and were diverted (1) or augmented (3) before BNR. Three patients failed after BNR to achieve continence, 2 due to incontinence alone, treated with diversion (1) or augmentation (1), and 1 who had hydronephrosis due to obstruction and poor compliance and needed augmentation.

Continence

Of 54 patients assessed for continence after BNR, 31 are dry both day and night, 11 are dry in the daytime only, 4 are dry with CIC (2 after bladder augmentation), 4 have a dry interval which is expected to improve, while 1 is wet, 1 is diverted and 2 are lost to followup. Thus the continence rate after BNR to date is 46 of 54 or 85.2 percent (Table 5). Overall, in the evaluable group, 49 of 57 are socially continent (86 percent) including 3 who became continent without BNR.

When considering those patients who had surgery attempting to produce functional bladder control, 45 of 57 are voiding urethrally and are socially continent, for a continence rate of 78.9 percent. The remaining 12 include 4 who are dry on CIC (2 after augmentation), 5 who do not have a satisfactory dry interval yet, 1 with diversion and 2 lost to follow-up.

Daytime continence occurred at a mean of 2.12 years (range 0.33 to 5.33) after BNR in 42 patients while night-time control occurred at a mean of 4 years (range 0.33 to 10.25) after BNR in 31 patients.

Osteotomy and Continence

Five patients closed without osteotomy are awaiting BNR. Of the remaining 19 patients, 14 are dry day and night (2 without BNR), 4 are dry during the day and 1 is lost to follow-up. This gives a rate of social continence of 95 percent (Table 6).

Table 5. Continence after BNR

Total BNR	58	
Dry—day and night	31	
Dry—day	11	
Dry—CIC	4	(2 after augmentation)
Dry interval	4	(Between 1 and 2 hours)
Recent BNR	4	(Less than 6 months)
Other	4	(Lost to f/u-2, Diverted-1, Wet-1)
Social continence	46 / 54	(85.2 percent)

Thirty-six patients who had osteotomy at primary closure have subsequently had BNR. Thirty of the 36 are socially continent for a rate of 83 percent (Table 6).

Selection regarding bladder size and condition, pelvic malleability and age at presentation make the group with and without osteotomy difficult to compare.

DISCUSSION

The goals of operative treatment in classic bladder exstrophy are to obtain a secure closure of the bladder and abdominal wall, to provide urinary continence and preserve renal function, and to reconstruct functional and cosmetically acceptable external genitalia.[11] This study chronicles the success obtained in achieving these goals, by treating the classic exstrophy bladder with a staged functional closure. Such an approach allows the imperfections in surgical results to be rectified before the succeeding stages. Earlier reports from this institution have indicated a high percentage of continence in our patients treated in this fashion.[12,13] The results of this review of our 21-year experience, in patients treated from birth at the Johns Hopkins Hospital, highlight the continuation of successful outcomes, thus attesting to the durability of this approach.

All exstrophic bladders should be closed and subsequently reconstructed in a staged manner. Evaluation as soon as possible after birth, with immediate bladder closure to preserve the bladder mucosa and prevent scarring secondary to exposure, helps to achieve optimal results. With prenatal ultrasound diagnosis,[14] it is possible to counsel the prospective parents and make appropriate preparations to avoid delays in initial management, especially when referral to tertiary care centers is involved. The initial closure is crucial in determining eventual urinary continence,[13] and is facilitated by osteotomies.[4,15] Very few

Table 6. Osteotomy and continence

No osteotomy		Osteotomy
19	Total number	36
12	Dry—day and night	18
4	Dry—day	7
—	CIC	4
—	Wet / diverted	2
1	Lost to follow-up	1
—	Recent BNR	3
2	Dry without BNR	1
18 (95 percent)	Socially continent	30 (83 percent)

bladders, if any, need to be excluded as unsuitable for primary closure; in this series, only 1 of 88 was treated with cystectomy and diversion and that too, due to parental choice, as mentioned above.

Paraexstrophy skin flaps have been used in 41 of the 63 boys in this group, while creating a new epispadiac meatus at the base of the pubis, after dividing the urethral plate. This has not been needed in any of the girls. Due to a high incidence of complications, especially strictures, requiring multiple corrective procedures, we rarely, if ever, use this technique at present. Further, the Cantwell-Ransley epispadias repair that follows, obviates the need for paraexstrophy flaps.

Bilateral osteotomies enable the mobilization and approximation of the widely separated halves of the pelvic ring, providing an obvious ease of abdominal wall closure. The deficiencies of the bony pelvis in classic bladder exstrophy,[16] including external rotation of the posterior pelvis, retroverted acetabula, additional external rotation of the pubic rami with shortening, and progressive diastasis of the symphysis pubis are best addressed by osteotomies. This also helps to bring the levators and the puborectalis sling together, around the urethra and to position the posterior urethra and bladder neck within the true pelvis (Figure 1), contributing to later continence. All of this may be accomplished in the neonate with a malleable pelvis without an osteotomy, but in the older infant or with wide diastasis of the symphysis pubis, pelvic osteotomy usually becomes necessary. This patient selection makes direct comparison of the final outcomes between the 2 groups difficult (Table 6). A posterior urethra superficially positioned between widely separated pubic bones has proven to be a severe obstacle in the attempt to gain continence in other patients, not in this series.

Postoperative immobilization is integral to the promotion of healing and prevention of dehiscence. This is accomplished effectively with an external fixator (Figure 4), that can be adjusted for optimal position, combined with modified Buck's traction for a period of 3–4 weeks, while children treated without an osteotomy are placed in modified Bryant's traction (Figure 3). Involvement of the parents, family members, supportive nursing care and adequate sedation ensures the comfort of the child in this frustrating period of immobilization, especially so for the older ones. External plaster fixation or "mummy-wrapping" on an outpatient basis has not provided adequate immobilization, as judged in other patients referred with partial or complete dehiscence.[3]

Ureteral stents, suprapubic catheters and drains are brought out through the neo-umbilicus or abdominal wall, avoiding the urethra. This prevents pressure necrosis and maceration of the wound edges due to urine leakage or friction caused by the catheters. Intravenous antibiotics started preoperatively are continued for 1 to 2 weeks, followed by suppressive oral doses until effective spontaneous bladder drainage is established. Regular sonograms detect obstruction to upper tracts, which if treated early, may not preclude continuation of the staged reconstruction. In this series, 2 of 86 developed hydronephrosis after initial closure and were treated with removal of the offending non-absorbable inter-symphyseal suture and ureteral reimplantation in 1 and cutaneous diversion in the other (performed at an outside institution).

Epispadias repair around age 1 to 2 provides some outlet resistance and allows the bladder to cycle itself, promoting growth.[8] Testosterone stimulation at 5 and 2 weeks prior to the procedure improves the vascularity and bulk of the tissues. At present we use a modified Cantwell-Ransley type of repair, which gives good cosmesis with lower incidence of fistulas.[10]

BNR is carried out when the child is mature enough and has developed a minimum of 60 cc of bladder capacity, as shown by cystogram under general anesthesia, typically at

4 to 5 years of age. We currently use a modified Young-Dees-Leadbetter procedure combined with a Marshall-Marchetti-Krantz bladder neck suspension (Figures 5–7), along with cephalotrigonal (Figure 8) or transtrigonal reimplantation to correct the associated reflux.[11,17] Postoperative urinary retention, seen in 12 of the 58 patients in this series resolved with a further period of catheterization or urethral dilation. Of 2 who developed hydronephrosis, 1 resolved with CIC while the other required bladder augmentation. Bladder stones formed in 7 patients, 6 of whom required surgical therapy. One patient had renal stones.

Urinary continence with a long dry interval, after such complex reconstructions, does take time and it is important that the patient and parents have realistic expectations. Bladder size and urodynamic evaluation at specific times before BNR and many months after bladder neck revisions may be very discouraging. However if the urine can be kept free of infection and if a dry interval has been achieved, between near complete evacuations, the bladder will increase in size and adverse urodynamic findings will disappear. The eventual achievement of full day and night control, even though considerably delayed, reduces the problem of infection, mucus accumulation, stone formation and possibly the eventual occurrence of malignant change.

With staged reconstruction, we regarded a complete success as retaining the native exstrophic bladder without renal compromise, urethral voiding, and continence without augmentation or CIC. In this series, 78.9 percent meet the above criteria.

As this successfully reconstructed exstrophic population enters adolescence and young adulthood, other pertinent issues will need to be addressed, including psychological health, sexuality, and fertility. Counseling by trained professionals with a special interest and insight into the problems encountered by these youngsters has proved helpful in some of our patients. Support groups that include parents and older patients are invaluable in addressing such issues.

The long-term potential for malignancy in the functionally reconstructed bladder can not yet be compared with the untreated exstrophy. A program of long-term surveillance with ultrasound, urinalysis and cytology must be instituted regardless of the apparent success of reconstruction.

ACKNOWLEDGMENT

The material forming the substance of this review, including the tables, has been submitted and accepted for publication in the Journal of Urology.

REFERENCES

1. Campbell, M: Embryology and anomalies of urogenital tract. In Clinical Pediatric Urology. Edited by Campbell, M. Philadelphia: W.B. Saunders Co., chapt. 3, p. 264, 1951.
2. Schultz, W. G.: Plastic repair of exstrophy of bladder combined with bilateral osteotomy of ilia. J. Urol., **79:** 453, 1958.
3. Gearhart, J. P., Ben-Chaim, J., Sciortino, C., Sponseller, P. D. and Jeffs, R. D.: The multiple reoperative bladder exstrophy closure: What affects the potential of the bladder? Urology, **47:** 240–243, 1996.
4. Jeffs, R. D., Guice, S. L. and Oesch, I.: The factors in successful exstrophy closure. J. Urol., **127:** 974, 1982.
5. Ansell, J. S.: Surgical treatment of exstrophy of the bladder with emphasis on neonatal primary closure: Personal experience with 28 consecutive cases treated at the University of Washington Hospitals from 1962 to 1977: techniques and results. J. Urol., **121:** 650, 1979.

6. Megalli, M. and Lattimer, J. K.: Review of the management of 140 cases of exstrophy of the bladder. J. Urol., **109:** 246, 1973.
7. Culp, O. S.: Treatment of epispadias with and without urinary incontinence: experience with 46 patients. J. Urol., **109:** 120, 1973.
8. Gearhart, J. P. and Jeffs, R. D.: Bladder exstrophy: Increase in capacity following epispadias repair. J. Urol., **142:** 525, 1989.
9. Marshall, V. F., Marchetti, A. A., and Krantz, K. E.: The correction of stress incontinence by simple vesicourethral suspension. Surg. Gynecol. Obstet., **88:** 509, 1949.
10. Gearhart, J. P., Sciortino, C., Ben-Chaim, J., Peppas, D. S. and Jeffs, R. D.: The Cantwell-Ransley epispadias repair in exstrophy and epispadias: Lessons learned. Urology, **46(1):** 92, 1995.
11. Gearhart, J. P. and Jeffs, R. D.: Exstrophy of the bladder, epispadias, and other bladder anomalies. In Campbell's Urology, 6th ed. Edited by P. C. Walsh, A. B. Retik, T. A. Stamey and E. D. Vaughan, Jr. Philadelphia: W. B. Saunders Co., vol. 2, chapt. 46, pp. 1772–1821, 1992.
12. Lepor, H. and Jeffs, R. D.: Primary bladder closure and bladder neck reconstruction in classical bladder exstrophy. J. Urol., **130:** 1142, 1983.
13. Oesterling, J. E. and Jeffs, R. D.: The importance of a successful initial bladder closure in the surgical management of classical bladder exstrophy: Analysis of 144 patients treated at The Johns Hopkins Hospital between 1975 and 1985. J. Urol., **137:** 258, 1987.
14. Gearhart, J. P., Ben-Chaim, J., Jeffs, R. D. and Sanders, R. C.: Criteria for the prenatal diagnosis of classic bladder exstrophy. Obstet. Gynecol., **85(6):** 961, 1995.
15. Gearhart, J. P., Forschner, D. C., Jeffs, R. D., Ben-Chaim, J. and Sponseller, P. D.: A combined vertical and horizontal pelvic osteotomy approach for primary and secondary repair of bladder exstrophy. J. Urol., **155:** 689, 1996.
16. Sponseller, P. D., Bisson, L. J., Gearhart, J. P., Jeffs, R. D., Magid, D. and Fishman, E.: The anatomy of the pelvis in the exstrophy complex. J. Bone and Joint Surg., **77-A(2):** 177, 1995.
17. Canning, D. A., Gearhart, J. P., Peppas, D. S. and Jeffs, R. D.: The cephalotrigonal reimplant in bladder neck reconstruction for patients with exstrophy or epispadias. J. Urol., **150:** 156, 1993.

25

PSYCHOLOGICAL ASPECTS OF BLADDER EXSTROPHY

Vulnerabilities, Risks, and Protective Factors from a Longitudinal Perspective

William G. Reiner

Division of Child and Adolescent Psychiatry
Johns Hopkins Hospital
600 N. Wolfe Street
Baltimore, Maryland 21287-3325

The dramatic transformation in the medical-surgical approaches to bladder exstrophy in the last three decades as well as in the results for exstrophy patients has not been legitimized by ongoing and concomitant studies of the psychological ramifications or patient adaptation. Such data that has been published is hampered by patient selection bias and an absence of longitudinal and systematic research. Because of recurrent anecdotal observations of psychological distress and problems with social adaptation in exstrophy patients, we initiated a set of longitudinal studies at the Johns Hopkins Hospital in 1993 looking at the psychosocial and psychosexual development in exstrophy patients and the implications for patients and family alike.

Ongoing studies following in excess of 100 children, adolescents, and young adults (from birth to 26 years) involve repeated cross-sectional assessments utilizing standardized and specially developed interview schedules of development, behavior, attitudes, adaptation, and activities. Social competence is assessed at a number of developmental ages (stages) as well, looking at academic and peer-relation performance and social activities. Psychosexual development is similarly assessed at numerous ages (stages). The subjects are screened for psychiatric illness as well. Self-image, body-image, and illness-adaptation and awareness are evaluated. Family structure, parental function, and sibling development are also areas of focus.

Preliminary results and analysis of data implies that exstrophy patients are resilient and adaptable overall, yet are vulnerable to major developmental hurdles and are more sensitive to many typical developmental stage-related obstacles. Developmental problem areas include peer relations in the pre-school to early primary school age group, social and

The Exstrophy–Epispadias Complex, edited by Gearhart and Mathews.
Kluwer Academic / Plenum Publishers, New York, 1999.

academic competence in the early middle school age, and psychosexual developmental problems and sexual dysfunction throughout adolescence. Adolescent sexual problems often have been previously unrecognized. Projected life span cannot yet be predicted, although one might predict an average at least into the fifth or sixth decades.

While analysis of sibling data is incomplete for any comments, early analysis of data from parents indicates major hurdles at the time of diagnosis and initial surgery, with any major surgical complications, when the child enters school, and when the child enters middle school. Sibling (family) issues are diverse yet involve parental absence most consistently. Parents deal with expressed and unrecognized guilt, with anger, with financial stress, and with partner-separation issues as well as with growing dependence upon—and, therefore, sometimes resentment of—the pediatric urologist and other members of the treatment team. Parents often focus their concerns appropriately enough on their child's medical and social condition, prognosis, and treatment but often to the neglect or exclusion of primary family and parent-relationship issues.

Unpredictability needs to be a primary concern for the child's medical life as well as for the family's social life. Hospitalization and medical requirements generally preclude a structured and consistent environment for the child. This heightens the child's risks for psychological maladaptation. In addition, there appears to be an increased risk of psychiatric illness in the face of vulnerability. In other words, with genetic (familial) illness-potential or with personality vulnerability, a child is likely to manifest psychiatric or social phenotypical responses at an earlier age and with more severity or difficulty than might otherwise be predicted. Thus, anxiety and stress responses, mood disorders, behavioral problems, and abnormal illness behaviors can be quite problematic, impacting the overall medical as well as social environment of the child.

Additionally, complications—predictable or otherwise—impact the child's growth and development, social environment, and self-esteem. Incontinence, for example, may be mild, moderate, or severe as a medical-surgical problem; yet the impact on the child or on the parents and the parenting may be pronounced. Demoralization is a common consequence of long term incontinence, beginning at least in the early school years. It can have a significant impact on the child's world—so much so that it may be wise to give serious consideration to anti-incontinence surgical approaches at the earliest feasible time in spite of the lack of data on outcome-impact. Similarly, other medical and surgical decisions should take into account the available—or sometimes even the predicted—social/psychological ramifications.

Psychosexual implications and ramifications of bladder exstrophy and its surgical repair may be profound. In a study of 16 consecutive adolescent males (ages 11 to 20), psychosexual developmental problems and dysfunction were universal. Adolescent males were 2 to 4 years delayed in psychosexual development and activity compared to their peers. Data on adolescent females is at present incomplete. Nevertheless, it is apparent that for adolescents with exstrophy, genital appearance and genital perception—that is, the sense of genital self-image—and body image are areas of major concern.

Observed resilience in this patient population seems to be accentuated by good parental relations, good doctor-patient-parent relations, and a good psychiatric liaison. The psychiatrist's assessing anxiety and anxiety-potential and providing interventions prior to surgical procedures seems to benefit patient and parents alike. Similarly, parent and patient education is not only useful medically but appears to aid in overall acceptance of individual problems as well as complications. The pediatric urologist may have to instruct the child psychiatrist in the medical-surgical aspects of exstrophy.

Education and counseling have been beneficial in early interventions with mid-adolescents and older adolescents. Acceptance of sexual deficiencies—real or perceived—seems to be the rule in those adolescents who have had a psychosexual assessment and have received sexual counseling; when those interventions are provided at early (appropriate) ages, anxiety decreases and the potential for mature, age-appropriate intimacy increases. Such interventions must not be static; that is, as psychosexual development evolves, the needs of the child will arise both individually and according to developmental stage. General education plus specific topics peculiar to exstrophy boys or girls is a valuable adjunct to medical treatment and to social development. The child psychiatrist knowledgeable about exstrophy can assess the patient for psychiatric illness or vulnerability while providing appropriate education for child or parents.

Finally, preliminary data from our study of the seven early to mid-adolescent children in the cloacal exstrophy population who were sex-reassigned at birth implies that maleness *may* be programmed into the brain by prenatal exposure to androgens, at least in some children. Other studies indicate a growing awareness that the differences in the male and female brains are at least partly structural. Until complete data are available, it would seem prudent to reevaluate the practice of sex-reassignment. The treatment team consisting of the pediatric urologist, pediatric endocrinologist, pediatric psychiatrist, and the primary pediatrician might review the available literature and provide a complete summary to the parents. Outside consultation by an expert in the area of gender identity development might be wise as well. Recognizing that this is an area beset by inadequate scientific data, *one* member of the team should be identified to make recommendations in order to provide a unified approach. A decision must be made, to be sure; yet there need be no overriding concern for urgency—that is, in a child with such major anomalies care must be taken to allow the parents to consider all the information available.

In summary, the complexities inherent to a congenital major malformation such as the exstrophy complex intimates the benefits of a concomitant medical and psychological approach to the care of these children. An important recommendation to the treatment team is the inclusion of a child psychiatrist. Psychiatric involvement should begin with the parents at the time of diagnosis and should continue as appropriate during decision making, at the time of surgical complications, and whenever the parent or the child demonstrate a need for help in dealing with the implications of the child's condition.

26

THE OUTCOME OF PATIENTS WITH CLASSIC BLADDER EXSTROPHY IN ADULT LIFE

Jacob Ben-Chaim,[1] Robert D. Jeffs,[2] William G. Reiner,[1] and John P. Gearhart[1*]

[1]The Division of Pediatric Urology
James Buchanan Brady Urological Institute
Department of Urology
[2]Division of Child Psychiatry
The Johns Hopkins Hospital, and
Johns Hopkins University School of Medicine
Baltimore, Maryland

ABSTRACT

Objectives

To determine the outcome of twenty adult bladder exstrophy patients regarding urinary continence, sexual function, fertility and psychosocial integration.

Materials and Methods

Twenty patients (16 males, 4 females) filled an anonymous questionnaire and their charts were reviewed for medical history.

Results

Nine patients are voiding spontaneously, with dry interval of 2 hours and more. Five patients are dry on clean intermittent catheterization. Three patients who had ureterosigmoidostomy are dry for more than 4 hours, and 3 had an incontinent stoma. Six patients are married (4 male, 2 female), of them 3 patients (2 males) had total of 7 children. Ten patients of 16 males (63%) reported to ejaculate a few cc in volume, 3 patients ejaculate only a few drops and 3 have no ejaculation (2 of whom underwent cystectomy). Semen

* Contact author: John P. Gearhart M.D., Marburg 149, The James Buchanan Brady Urological Institute, The Johns Hopkins Hospital, Baltimore, Maryland 21287-2101

analysis was done in 4 patients with the average volume of ejaculate being 0.4 cc (range 0.2–1 cc) 3 of whom had azoospermia and 1 oligospermia. All females reported regular and normal periods. Fifteen patients experienced normal erections which were described as satisfactory in 8 and not satisfactory in 6 because of a small penis and in 1 because of dorsal chordee. Twelve of the 16 males (75%) experienced satisfactory orgasms while 10 patients had participated in sexual intercourse with full partner satisfaction in nine. Half of the males and all females describe their intimate relationships as serious and long term. Fifteen of 20 patients (75%) achieve high level of education.

Conclusions

Bladder exstrophy patients generally achieve good results, however the fertility of most of the males is in doubt.

The staged reconstruction of bladder exstrophy generally achieves good results with urinary continence and cosmetically pleasing genitalia.[1,2] However, by the time of puberty through adulthood, other factors gain more importance for patients with bladder exstrophy such as self esteem, sexuality, sexual function, fertility, and psychosocial integration. In females, reconstruction of the external genitalia is usually uncomplicated and the fertility is normal.[3] Reconstruction of the male external genitalia and urethra is more complicated and in the past was more prone to complications. Since exstrophy patients have undergone multiple reconstructive surgeries of the urinary tract and many had recurrent urinary tract infections their fertility is at risk. Also, these patients have a relatively smaller penis. These factors may adversely affect the long term outcome of these patients. Since classic bladder exstrophy currently can be diagnosed prenatally, the outcome of patients with classic bladder exstrophy in the adult life is of great importance for prenatal counseling.[4] The authors, report the outcome of 20 adult exstrophy patients who have been treated in our institution.

MATERIAL AND METHODS

The charts of twenty adult bladder exstrophy patients mean age of 24 years (range 18–42 years), who have received treatment and are followed at our institute (16 males, 4 females) were reviewed for medical history. Six patients have been treated primarily from birth in our institute and 14 patients were referred to us after initial treatment elsewhere. In addition all of these patients filled an anonymous questionnaire in order to objectively assess the outcome regarding urinary continence, sexual function, fertility and psychosocial integration. This questionnaire was sent to total of 48 patients, however only 20 patients have responded and all are included in this study. Fourteen patients underwent staged primary exstrophy reconstruction that included bladder closure soon after birth, at age 2–3 years a modified Young epispadias repair and at age 4–5 Young Dees Leadbetter bladder neck reconstruction. Three patients underwent ureterosigmoidostomy, and 3 patients underwent ileal conduit urinary diversion (2 of whom underwent cystectomy).

RESULTS

Nine patients void spontaneously, 6 of whom are dry for 4 hours or more and 3 patients are dry for 2 hours. Five patients are on clean intermittent catheterization, 3 of whom are dry for 4 hours or more while 3 are dry for 2–3 hours. Three patients who have a ureterosigmoidostomy are dry for more than 4 hours, and 3 have an incontinent ileal stoma. In addition, four patients reported that they have significant nighttime wetting (3 of whom are on CIC and 1 voids spontaneously). Ten patients have had episodes of pyelonephritis while 3 patients have had attacks of epididymitis. Renal function was normal in 19 patients.

Six patients are married (4 male, 2 female) at age range of 24–42 (mean 30) years. Two males have fathered 2 and 3 children respectively and 1 female has 2 children. Ten of 16 males (63%) report an ejaculate of a few cc's in volume, 3 patients ejaculate only a few drops and 3 have had no ejaculation (2 of whom underwent cystectomy), while a clear history of retrograde ejaculation was reported in 6 patients who had minimal emission and noticed that seminal fluid present in the first urine after the emission. Semen analysis was done in 4 patients (2 of whom had urinary diversion) with an average ejaculate volume of 0.4 cc (range 0.2–1 cc) 3 of whom had azoospermia and 1 oligospermia. One of the patients with azoospermia underwent a bilateral vasogram which revealed bilateral complete obstruction of the ejaculatory ducts. All females reported normal and regular and periods.

Fifteen patients experienced normal erections which were described as satisfactory in 8 and not satisfactory in 7, (in 6 because of a small penis and in 1 because of persistent dorsal chordee). Twelve of the 16 males (75%) experienced satisfactory orgasms and 10 patients had participated in sexual intercourse with full partner satisfaction in nine. All of the females have been involved in sexual intercourse and 3 have experienced orgasms which were described as satisfactory in 2 patients with full partner satisfaction in all. One patient had uterine prolapse that required reduction prior to intercourse, while another patient had a dry vagina that required lubrication prior to intercourse.

Half of the males describe their intimate relationship as random and short term, while all females describe their intimate relationship as serious and long term. Ten patients are in college or plan to go to college, 5 patients study in the university (2 of them are medical students), 3 have graduated high school, one finished 5th grade and is a factory worker and one patient finished 8th grade and is unemployed. In addition the patients rank their satisfaction from the reconstructive surgery from 1 to 5 (1 being not satisfactory at all and 5 — very satisfactory) the range was 3 to 5 (mean of 3.83).

DISCUSSION

Both male and female bladder exstrophy patients usually achieve normal sexual function with generally full patient and partner satisfaction, but it is unclear how realistic the responses are about partner satisfaction, since the partners were not asked directly. Nevertheless, half of the males are bothered by their relatively small penis. This group of patients have good psychosocial integration and most of them achieve high levels of education. However, this may represent a selection bias in the study since the more educated patients may tend to cooperate with this type of study.

According to this study it appears that most male exstrophy patients may have a fertility problem, since only 12% of the males fathered children and thus seem to be infertile. However, most of these patients are quite young and have not reached the usual age of marriage and reproductive and 50% of the married patient had children. The infertility in the male exstrophy patient may be secondary to recurrent and chronic urinary tract infections and epididymitis, retrograde ejaculation, and previous surgical procedures in the prostatic urethral area. In fact, each stage in the bladder exstrophy reconstruction may harm the delicate ejaculatory ducts orifices in the prostatic urethra and result in ejaculatory duct obstruction and azoospermia as was documented in one of these patients. Moreover, in cases who have had cystectomy and urinary diversion a higher risk of infertility exists due to damage to the genital tract occurring during cystectomy. Therefore, when urinary diversion is required, it is generally recommended to perform subtotal cystectomy leaving of a small pouch for collection of the ejaculate and to carefully preserve the vas deferens on each side.

Stein et al.[5] found similar results in 45 adult patients who had undergone ureterosigmoidostomy. Only 5 patients who had undergone no genital reconstruction had normal ejaculation and only 2 of 31 men had fathered children. An abnormal semen analysis was found in all 5 patients. Woodhouse et al.[6] reviewed 74 adult (52 males) exstrophy patients and found good sexual function in all. However, only 6 were found to be fertile, 5 of whom had undergone early subtotal cystectomy and diversion. Feitz et al (7) looked specifically at the psychosexual and socioeconomic development in 22 patients most of whom underwent urinary diversion. They found that all have a positive attitude towards life. However, only 7 of the 10 males described their erections as satisfactory and 4 complained of retrograde ejaculation. Also, penile length and chordee was found to be the most embarrassing issue for the young men in their study.

Males appear to be vulnerable to intimacy problems in male-female relationships, although the effects of exstrophy on psychosexual development is unclear at present. Moreover, the fertility problems experienced by most of the male patients may negatively impact self-esteem and may lead to further deterioration in relationships.

Recently, we changed the technique of epispadias repair to the Cantwell-Ransley technique which achieves ventral displacement of the urethra and meatus, better correction of the dorsal chordee and better cosmesis (8). Therefore, the authors believe that the outcome of male exstrophy patients will improve even more in the future.

These data clearly demonstrate that the long term results in exstrophy patients other than male infertility are quite acceptable. Therefore, when prenatal diagnosis of exstrophy is made, the parents can be confirmed that the long–term outcome for these children is optimistic.

In conclusion, bladder exstrophy patients achieve good results as far as continence is concerned, sexual function is fair in the male and good in the female; and psychosocial integration appears to be adequate. However, the fertility in most of the male patients is in doubt and psychosexual developmental issues may be more of a problem for the male than for the female. With current refinements in the treatment of bladder exstrophy and male infertility, the results are expected to improve in the future.

REFERENCES

1. Gearhart J.P. and Jeffs R.D.: Bladder Exstrophy: Increase in Capacity Following Epispadias Repair. J. Urol. 142:525, 1989.
2. Connor J.P., Hensle T.W., Lattimer J.K., Burbige K.A.: Long Term Follow-Up of 207 patients with Bladder Exstrophy: An Evolution of Treatment. J. Urol. 142:793, 1989.
3. Krisiloff M., Puchner P.J., Tretter W., Macfarlane M.T. and Lattimer J.K.: Pregnancy in Woman with Bladder Exstrophy. J. Urol. 119:478, 1978.
4. Gearhart J.P., Ben-Chaim J., Jeffs R.D. and Sanders R.: Criteria for prenatal diagnosis of bladder exstrophy. Obstet & Gynecol. 85(6) June 1995 (accepted for publication).
5. Stein R., Stockle M., Fisch M., Nakai H., Muller S.C. and Hohenfellner R.: The Fate of Adult Exstrophy Patients. J. Urol. 152:1413, 1994.

6. Woodhouse C.R.J., Ransley P.G. and Williams D.I.: The Patient with Exstrophy in Adult Life. Brit. J. Urol. 55:632, 1982.
7. Feitz W.F.J., Van Grunsven E.J.K.J.E.M., Froeling F.M.J.A. and de Vries J.D.M.: Outcome Analysis of the Psychosexual and Socioeconomical Development of Adult Patients Born with Bladder Exstrophy. J. Urol. 152:1417, 1994.
8. Gearhart J.P., Sciortino C., Ben-Chaim J., Peppas D.S. and Jeffs R.D.: The Cantwell-Ransley Epispadias Repair: Lessons Learned. "Urology" 46:92, 1995.

LONG TERM RESULTS OF BLADDER EXSTROPHY

The London Experience

C. R. J. Woodhouse

The Institute of Urology
The Hospital for Sick Children
Great Ormond Street, London, United Kingdom

Classical exstrophy has been recognised for about 2000 years and has been the subject of voluminous surgical literature since the middle of the last century. Contemporary surgery for this condition has developed on the basis of ideas formed nearly a century ago. The posterior pelvic osteotomy and bladder reconstruction used by Trendelenburg in Leipzig at the turn of the century is similar to that used today. He also mentioned that Thiersch had implanted a 'special mechanical contrivance' to maintain continence after reconstruction which sounds like a fixed pressure artificial sphincter.[1] The photographs in Trendelenburg's paper show a very acceptable cosmetic result and a young boy able to void with a most impressive stream from very normal looking penis. As so often, it is not the surgical ideas that have changed with time, but the materials and drugs necessary to support those ideas.

As has been shown in previous chapters, the result of exstrophy reconstruction, both functional and cosmetic, have improved enormously in the last 20 years. The writer on long term follow up is, therefore, in the invidious position of describing results that now would be unacceptable. Nonetheless, it is only by looking at the long term results that improvements can be made in neonatal management. It is striking that problems once common in our adolescent clinic, such as severe penile chordee, are now rare.

MORTALITY

Exstrophy is not, in itself, a lethal abnormality but the complications and treatment often were. In 1926 it was recorded that 72% of 66 children were alive but nearly half

The Exstrophy–Epispadias Complex, edited by Gearhart and Mathews.
Kluwer Academic / Plenum Publishers, New York, 1999.

Table 1. Causes of, and ages at death, in 9 of 101 cases of exstrophy[4]

Cause of death	Number	Age
Urological		
Renal failure*	2	8 months 12 years
Ca. bladder*	1	29 years
Non-urological		
Intestinal obstruction	2	7 years 10 years
Inhalation of vomit	1	5 years
Carcinoma of colon*	1	30 years
Poliomyelitis	1	3 months
Unknown (? accident)	1	17 years

*Deaths related to exstrophy and its treatment.

were under 10 years old. Historically, 50% of patients were said to be dead by 10 and 67% by 20 years old. There was long term follow-up on 20 patients from 30 to 73 years old.[2]

In the post-war period in New York, only 5 deaths were recorded from 1945 to 1970, 3 of which were in patients with ureterosigmoidostomies. Obviously the follow-up was short in some cases and 39 were lost altogether so the overall mortality may have been higher.[3]

Our own adult series published in 1983, included 103 cases with a minimum follow-up of 18 years with only 10 patients lost before puberty. There were 9 deaths and as their follow-up has continued no more have died. Deaths occurred at ages from 3 months to 30 years but only 4 could be attributed to the exstrophy and its treatment (Table 1).[4] Death from exstrophy alone now would be most uncommon.

Late deaths show a similar pattern. In a group of 87 patients presenting between 1912 and 1935 and reviewed in 1974, the 25 deaths could all be attributed to the urinary diversion by ureterosigmoidostomy which was the standard management or to non-urinary tract causes.[5]

As there seems to be some renewed enthusiasm for ureterosigmoidostomy (see chapter 16), it is appropriate to consider the cancer risk. The increased risk of carcinoma in patients with ureterosigmoidostomy has been variously estimated at 100 to 7000 times that in the normal population.[6]

In 103 adult exstrophy patients, Strachan has looked at the cancer incidence at a minimum follow-up of 20 years. He divided them into those who had ever had urinary and faecal streams mixed (at risk group — 42 patients) and those who had had an early diversion by colonic or ileal loop (non-risk group — 61 patients). There were no cancers in the non-risk group. In the 'at risk' group there were 4 cancers and 4 non-invasive neoplasms at the ureterosigmoidostomy. Two further patients in this group died of adenocarcinomas said to have arisen in the bladder. The mean age of the 'at risk' patients was 36.1 years and the cumulative risk of developing colonic neoplasia at that age is 6.1 per 100,000 for normal subjects. In the ureterosigmoid patients (excluding the bladder cancers) the incidence was 10,526 per 100,000. Therefore, in this relatively pure group, the risk is 1,726 times the expected incidence.[7]

When considering an individual patient, it must be remembered that the risk is a continuing one: in the 42 patients at risk, under personal surveillance, there has been one further ureterosigmoid adenoma and another bladder cancer. In this particular group 31% have developed neoplasms and presumably more will do so.

THE BLADDER

Although good functional results have been reported for bladder closure by several authors over the last 100 years, the percentage of fully continent exstrophy patients was small. Reviewing the literature over the previous 60 years Marshall and Muecke found that only 5% had a 'good' result and 14% a 'fair' result from bladder reconstruction. Their own series had a 20% 'good' result rate but follow-up was short.[8] In 1973, with experience of 140 patients, Megali and Lattimer recommended ileal conduit diversion to give the best social and psychological results with the least morbidity.[3]

The policy at Great Ormond Street at that time was to close only those bladders that appeared to have a reasonable size and to have good tissue. The decision was usually taken at 6 to 18 months of age. With this policy an attempt was made to close 62 of 103 bladders. Many techniques were tried but the outcome appeared to depend more on the original bladder size than anything else; performance of a pelvic osteotomy and age at surgery were irrelevant. Indeed, most of the patients who had an osteotomy had re-established a widely open symphysis by adult life without adverse effect on their continence. Completely reliable continence was achieved in 5 of 21 girls and 5 of 41 boys. One other girl had partial continence requiring the use of 1 pad per day. One other boy who later died, was enuretic though dry by day.[4] The greater success rate in girls has been noted by others and remains unexplained.

The continent exstrophy bladder in adult life has been a cause of concern. Four patients have now formed stones. When they are cystoscoped, the bladder looks like the inside of an egg shell even though it does not appear calcified on plain x-ray. I presume that fragments break off the wall from time to time and form the nucleus for a proper stone.

The long term bladder function of a cohort of patients born in the decade 1965 to 1974 has been reviewed in detail. There were 57 evaluable patients (two others had been lost to follow-up) of whom 25 underwent primary diversion and 32 were reconstructed. Of the reconstructions, 19 failed early and 13 had a working (unaugmented) bladder at the begining of adulthood.

Only three have maintained normal function in their third decade of life. The remainder have required augmentation, artificial sphincters or other reconstruction and/or needed to start intermittent self catheterisation.[9]

This depressing finding may be explained, in part, by the late age at which the original bladder closure was done and the lack of knowledge of the dynamic function of the bladder outlet in exstrophy[10]. Nonetheless, even the most modern bladder neck reconstruction does not create a working sphincter. The outlet resistance is, in fact, an obstruction. If the bladder empties by detrusor contraction, it is likely that its function will deteriorate with time, because of instability and loss of compliance.

FEMALE GENITALIA

Little attention has been paid to the female genitalia in exstrophy. The need to open the tight introitus and the high risk of uterine prolapse were recognised. Recent studies of the adult females have indicated that patients are concerned about the appearance of their genitalia and that reconstruction can give a near normal appearance.

A sagittal section of the pelvic organs is shown in Figure 1. It will be seen that the vagina, introitus and anus are each displaced anteriorly. The vagina is shorter than normal, seldom measuring more than five or six centimetres, but of normal calibre. The uterus

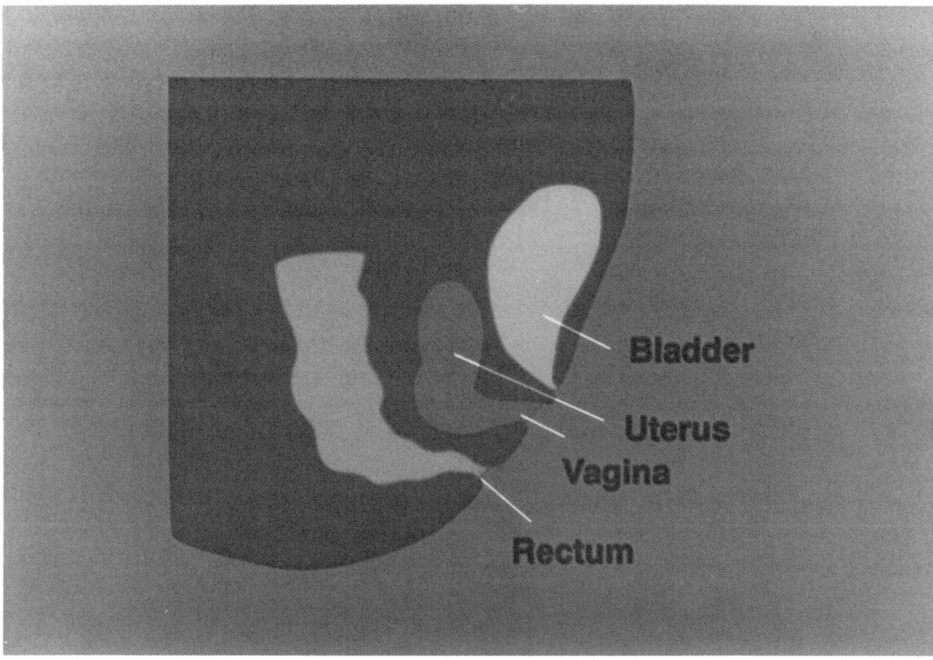

Figure 1. Diagram of the sagittal section of the female pelvis in classical exstrophy. Note the anterior and cranial displacement of the organs

which is in an almost normal anatomical position, enters superiorly so that the cervix is in the anterior vaginal wall. The ovaries and tubes are normal.[11]

Appearances in Adult Females

The labia and introitus appear to be structures of the lower anterior abdominal wall (Figure 2). The introitus is narrow (Figure 3). The narrowing is posterior and is a more substantial layer than the normal hymen, probably being a continuation of the posterior wall of the vagina. The anus is in the position of the normal vagina. The bifid and rudimentary labia can be seen. The pubic hair lies on either side of the mons as is seen in males (Figure 4).

Again, this abnormal appearance is variable and in some females, the labia are almost in the correct anatomical position. Those with a working bladder generally have a more normal appearance, otherwise, the original severity of the epispadias seems irrelevant.

Reconstruction

The object of the labial reconstruction is to unite the two halves of the clitoris (assuming that they can be identified) and to fuse the anterior ends of the labia to make a fourchette. It is not possible to move the vagina posteriorly to its usual anatomical position, but labial and pubic reconstruction does disguise the abnormality very well.[11]

In spite of the very narrow introitus (Figure 3), the vaginal calibre above is normal. The vagina is opened by doing an episiotomy posteriorly from the introitus until a two finger opening is created. It is usually possible to close the vaginal mucosa to the perineal skin directly. Sometimes a flap of mucosa from the medial aspect of the labia is needed on

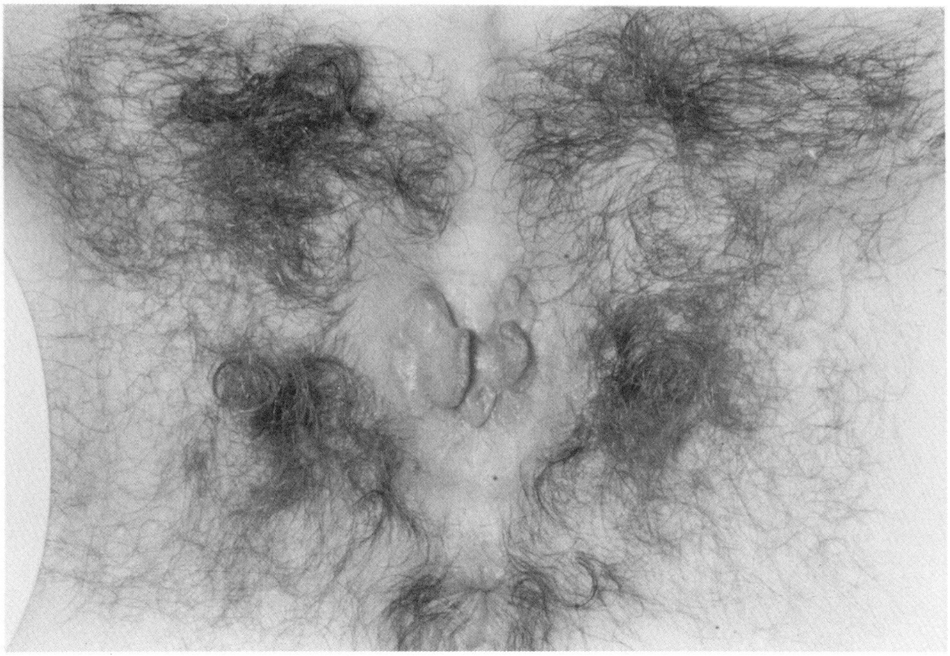

Figure 2. Clinical photograph to show the appearance of the vagina in an adult exstrophy patient.

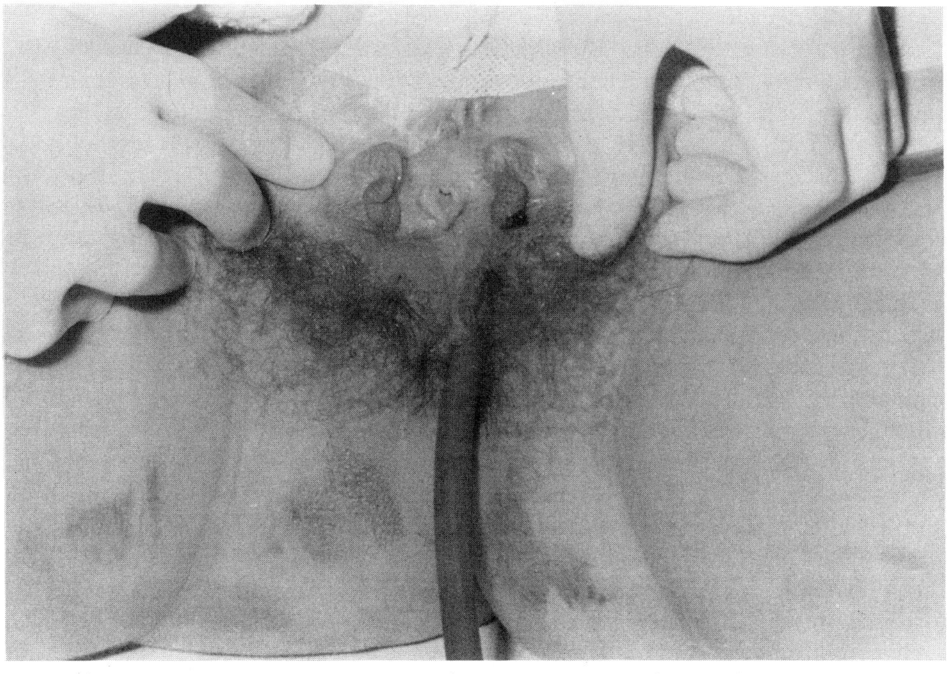

Figure 3. Clinical photograph showing the narrow introitus through which a probe could just be passed (the Foley catheter is in the anus). Note the bifid clitoris and labia minora.

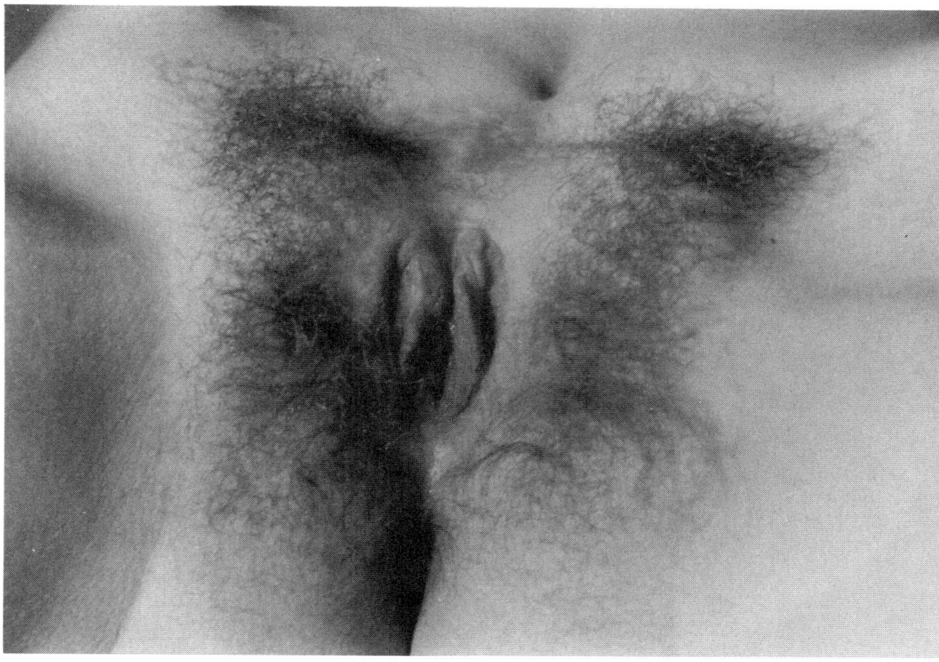

Figure 4. Clinical photograph of the distribution of pubic hair.

each side to close the defect. With experience I have found that the need for formal vaginoplasty for the introitus has reduced.

The Mons Pubis. In adult exstrophy patients at present, the pubic area is nearly always recessed from the uncorrected divarication of the pubic bones. The pubic hair lies on either side of the midline (Figure 4).

It is most important, either in infancy or in adolescence, to rotate hair bearing flaps of skin and fat to cover the mid line defect. The flaps may be based laterally or inferiorly.[11–13] If there is too little skin available, larger flaps can be created by the use of skin expanders.[14] The only problem with the use of tissue expanders (or using flaps under a little tension) is that the number of hair follicles is not increased. Thus, although good skin cover is achieved, the pubic triangle can look a little 'bald'.

The scarred and non-hairy skin from the midline is excised. In the female, this excision may be continuous with that from the anterior end of the labia. The flaps of hair bearing skin, including all of the subcutaneous fat are then mobilised and rotated to fill the defect (Figure 5). In females the defect left by re-positioning of the flaps can be primarily closed.

Contemporary techniques of primary reconstruction recognise the distribution of hair bearing skin and position it appropriately. Unfortunately, the benefit of such surgery may be lost if many revision operations cause unsightly mid-line scars later on.

As with males, there is a tendency to reconstruct the genitalia in infancy or childhood. Hohenfellner's group suggest opening of the introitus and uniting the labia at three or four years of age.[15] This allows easy drainage of secretions and, later, menstrual flow. The more natural appearance in the school years is welcome. When reviewed after puberty excellent results from early reconstruction were found in 21 of 25 girls. However, most of these patients had been diverted and did not have the sequence of revision operations that is so common after primary reconstruction.

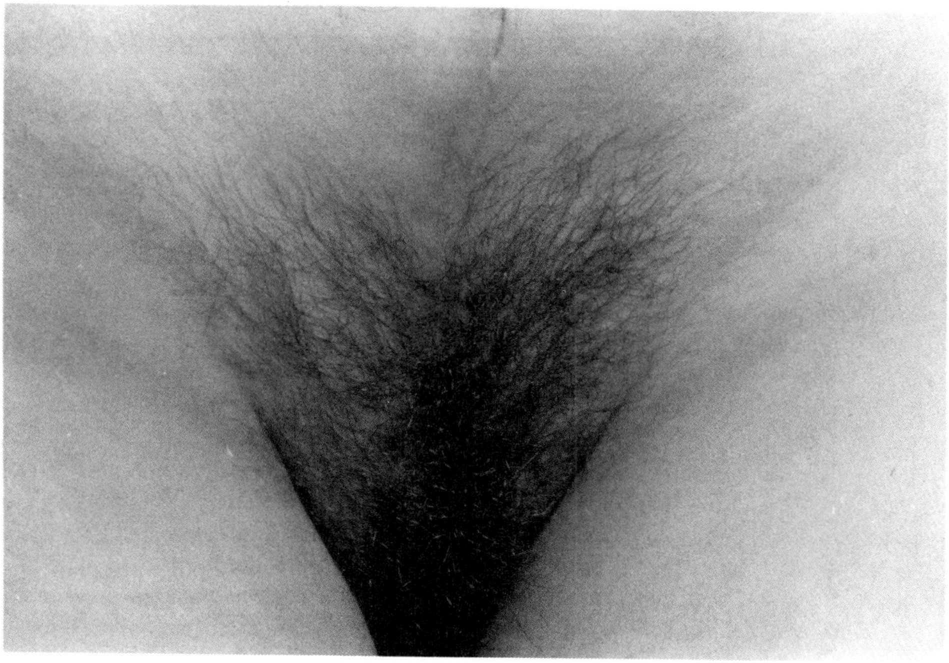

Figure 5. Post operative clinical photograph.

In my own series of adult females 28 of 42 have had a modified episiotomy and 2 a vaginoplasty to allow penetrative intercourse. Fifteen patients had a vulvoplasty and a monsplasty. The surgical results were good in all but one case. One patient had partial necrosis of the skin flaps for the monsplasty.[12]

Sexual Function

The vagina runs directly posteriorly so that when the patient is standing the vagina is almost parallel with the floor. It is also rather shorter than usual. These arrangements do not interfere with intercourse: all of 30 reconstructed women in my series have normal intercourse and none report any difficulties, while only four unreconstructed cases do so.

Orgasm is said to be normal in all patients.[16] All of 19 patients over 18 years old in Stein's series have intercourse but three consider it unpleasant. The source of unpleasantness is not given.[15]

Considering the fairly obvious abnormality of the vagina, both in position and appearance, it is surprising that more emotional problems are not seen. I am only aware of one patient in my series who has severe problems. Several females note that the rather broad perineum, that is a consequence of the outward rotation of the pelvis, makes the wearing of modern narrow crotched bathing costumes impractical.

Uterine Prolapse

The defective pelvic floor and poor uterine supports make prolapse common. It is a considerable problem to correct.[17,18] Seven of my patients have had a total procedentia, one of whom had never had intercourse or a pregnancy. The uterus should not be removed as it is the only solid organ in the pelvis that has any hope of holding up the pelvic floor.

For procedentia, the most successful procedure is the gortex wrap. The sacral promontory is exposed. A strip of gortex is sutured to the periosteum. The end is passed around the cervix through the base of the broad ligament and brought back to the sacrum.[19] This procedure has been successful in all cases with follow-up to a maximum of six years.

Hohenfellner advocates fixation of the uterus to the anterior abdominal in childhood. This is said to prevent prolapse but still allow normal pregnancy.[15] Two women were able to have normal pregnancies without prolapse, while one of two women who did not have a fixation had slight prolapse after delivery. This 'prophylactic surgery' may well be helpful. However, once prolapse has occurred, I have not found an anterior fixation to be an effective repair.

Fertility

The cervix enters the superior wall of the vagina and is very close to the introitus: this must contribute to the ease with which the girls get pregnant. It is essential to explain to the patients at an early stage that fertility is normal.

Pregnancy has a poor reputation in females with exstrophy probably because of the literature review from 1722 to 1958 by Clemston.[20] There were only 49 live births from 64 pregnancies and two maternal deaths.

Better obstetric care has improved the results. In a combined series of 22 patients there were 32 pregnancies. Two ended in spontaneous abortion and two in therapeutic abortion. There was only one obstetric catastrophe which was the intra uterine death of twins. There were 27 live births.[21]

Pregnancy and delivery are usually normal. Elective caesarean should be done in patients with a working bladder or with an artificial urinary sphincter (11 of the 32 deliveries mentioned above). The principle problem is subsequent uterine prolapse.

Inheritance

It has been reported from a postal survey of clinics throughout the world, that 1 in 70 offspring have exstrophy or epispadias.[22] This figure must represent the worst that can be expected.

Screening pregnancies for foetuses with exstrophy is difficult; advising the parents on whether or not to terminate an affected foetus is almost impossible. Foetal ultra sound at 18 to 20 weeks may show that the bladder does not fill. This, combined with a low umbilicus and ambiguous genitalia if the foetus is male, should raise the possibility of exstrophy. However, in a series of 30 antenatally diagnosed urological anomalies, five were thought to have an absent bladder but all had other major anomalies as well. On delivery or termination, none had exstrophy.[23]

In view of the difficulties with ultra sound diagnosis, foetoscopy seems justified for suspected cases. In two recent pregnancies, classical exstrophy has been confirmed on foetoscopy (Dhillon, personal communication 1997).

REFERENCES

1. Trendelenburg F: The treatment of ectopia vesicae. Ann Surg 1906, 44: 281–289.
2. Mayo C H, Hendricks W A: Exstrophy of the bladder. Surg Gyne Obstet 1926, 43: 129–134.

3. Megali M, Lattimer J K: Review of the management of 140 cases of exstrophy of the bladder. J Urol 1973, 109: 246–248.
4. Woodhouse C R J, Ransley P G, Williams D I: The patient with exstrophy in adult life. Brit J Urol 1983, 55: 632–635.
5. Segura J W Kelalis P P: Long term results of ureterosigmoidostomy in children with bladder exstrophy. J Urol 1975, 114: 138–140.
6. Stewart M: Urinary diversion and bowel cancer. Ann Royal Coll Surg (England) 1986, 68: 98–102.
7. Strachan J R, Woodhouse C R J: Malignancy in exstrophy patients following ureterosigmoidostomy. Brit J Surg 1991, 78: 1216–1218.Strachan
8. Marshall V F, Muecke E C: Functional closure of typical exstrophy of the bladder. J Urol 1970, 104: 205–212.
9. Woodhouse CRJ, Redgrave: Late failure of the reconstructed exstrophy bladder. Brit J Urol 1996, 77: 590–592.
10. Hollowell J G, Hill P D, Duffy P G, et al. A preliminary report: bladder function and dysfunction in exstrophy. Lancet 1991, 338: 926–928.
11. Woodhouse C R J, Hinsch R: The anatomy and reconstruction of the adult female genitalia in classical bladder exstrophy. Brit J Urol 1997, 79: in press.
12. Ship A G, Pelzer R H: Reconstruction of the female escutcheon in exstrophy of the bladder. Plast Reconst Surg 1972, 49: 643–646.
13. Marconi F, Messina P, Pavanello P, et al: Cosmetic reconstruction of the mons veneris and lower abdominal wall by skin expansion as the last stage of surgical treatment of bladder exstrophy. Plast Reconst Surg 1993, 91: 551–555.
14. Fid J F, Rosenberg P, Rothaus K, et al: The use of tissue expanders in the final reconstruction of infra pubic midline scar, mons pubis and vulva after bladder exstrophy repair. Urology 1993, 41: 426–430.
15. Stein R, Fisch M, Bauer H, et al: Operative reconstruction of the external and internal genitalia in female patients with bladder exstrophy or incontinent epispadias. J Urol 1995, 154: 1002–1007.
16. Krisiloff M, Puchner P J, Tretter W, Macfarlane M T, Lattimer J K: Pregnancy in women with exstrophy. J Urol 1978, 119:478–479.
17. Dewhurst J, Topliss P H, Shepherd J H: Ivalon sponge hysteropexy for genital prolapse in patients with bladder exstrophy. Brit J Obst Gynaecol 1980, 87: 67–69.
18. Blakeley C R, Mills W G: The obstetric and gynaecological complications of bladder exstrophy and epispadias. Brit J Obst Gynaecol 1981, 88: 167–173.
19. Farkas A G, Shepherd J H, Woodhouse C R J: Hysterosacropexy for uterine prolapse associated with urinary tract abnormalities. J Obstet Gynaecol 1993, 13: 358–360.
20. Clemston C A B: Ectopia vesicae and split pelvis. J Obstet Gynaecol 1958, 65: 973–981.
21. Woodhouse C R J: The sexual and reproductive consequences of congenital genitourinary anomalies. J Urol 1994, 152: 645–651.
22. Shapiro E, Lepor H, Jeffs R D: The inheritance of the exstrophy-epispadias complex. J Urol 1984, 132: 308–310.
23. Pocock R D, Witcombe J B, Andrews H S, Berry P J, Frank J D: The outcomes of antenatally diagnosed urological abnormalities. Brit J Urol 1985, 57: 788–792.

28

LONG TERM RESULTS OF BLADDER EXSTROPHY

The Mainz Experience

Raimund Stein,* Margit Fisch, and Rudolf Hohenfellner

Department of Urology
Johannes Gutenberg University, Medical School
Mainz, Germany

ABSTRACT

After primary bladder closure or urinary diversion, other factors apart from the reconstruction gain importance for individuals with the exstrophy-epispadias complex: social integration and, after reaching puberty, sexuality and fertility.

Between 1968 and 1994 115 patients with bladder exstrophy or incontinent epispadias underwent surgery at our institution. A total of 104 patients could be followed, 2 of whom died in the meantime.

94% of the patients with rectal reservoirs and 97% of those with a Mainz pouch I are continent day and night, whereas only 3 of 6 patients with a sling plasty (incontinent epispadias) or with primary bladder closure followed by a Young Dees procedure are continent. The upper urinary tract remained stable at the latest followup in between 95% (Mainz pouch) and 98% of the renal units (rectal reservoirs).

Of the 104 patients 48 attend school, 4 are in college, 42 have completed or are currently undergoing vocational training, 3 are unemployed, 1 lives in a therapeutic center, 4 attend kindergarten and 2 are younger than 4 years of age.

In females reconstruction of the external genitalia resulted in normal sexuality and fertility. Reconstruction was performed in 25 and fixation of the uterus in 13 to correct or prevent uterine prolapse. Of the 17 women older than 18 years of age with genital recon-

* Dr. med. Raimund Stein, Langenbeck Str. 1, 55131 Mainz, Germany, Tel. 0049 6131 177177, Fax: 0049 6131 232986 or 230462

The Exstrophy–Epispadias Complex, edited by Gearhart and Mathews.
Kluwer Academic / Plenum Publishers, New York, 1999.

struction, 16 are satisfied with the cosmetic result. All adults engage in sexual intercourse. Five women have delivered 7 children by Caesarean section.

In male patients cosmetic results were satisfactory, however, at the expense of the fertility. Penile deviation was present in 11 of the 32 patients with genital reconstruction, which is distressing in only 2. Thirty patients are satisfied with the cosmetic result. 9 males developed epididymitis, necessitating 2 orchiectomies and 3 vasectomies. No patient with reconstruction of the external genitalia ejaculates normally or has fathered children, whereas ejaculation was normal in 3 men who did not undergo genital reconstruction. Of these, 2 men have fathered 4 children.

Conclusion: In individuals with the bladder exstrophy-epispadias complex Education, Occupation and social development is normal. The cosmetic results achieved by genital reconstruction are satisfactory. In female patients, antefixation of the uterus should be performed before or together with an introitusplasty to prevent uterine prolapse. In male patients, however, surgery is performed at the expense of fertility.

INTRODUCTION

Most publications concerning the treatment of patients with bladder exstrophy or incontinent epispadias present operative techniques and continence rates. The effect of primary bladder closure or urinary diversion on social integration is often neglected in the literature.[1] The same holds true for reconstruction of the external and, in women, the internal genitalia. When the problems associated with urinary incontinence have been solved and the patients have reached puberty, social integration, sexuality and fertility become more significant to these patients.

These children have a normal life expectancy if renal function is normal. Continence is important for a good social integration.

The operative techniques for reconstruction of the external genitalia in females are not problematic, however, after correction of the very narrow introitus, uterine prolapse is not uncommon.[2–4] This complication can be prevented or corrected by fixation of the uterus.[1,5]

In male patients, the cosmetic results after one or more reconstructions of the penis are satisfactory.[6–9] However, there are only few reports in the literature concerning the consequences of genital reconstruction, especially with respect to fertility.[1,10,11]

The social integration of patients after urinary diversion, as well as their sexual behaviour and fertility, are presented. Possibilities and consequences of the operations, mainly performed during childhood, are discussed.

PATIENTS AND METHODS

Between 1968 and 1994, 95 patients with bladder exstrophy and 20 with incontinent epispadias were operated upon at our institution.

A total of 104 patients could be followed. Of the 115 patients, 10 were lost to follow-up and one patient refused participation. Two patients have died, one of liver failure as a result of alcohol toxicity, the other of metastases of a signet ring cell carcinoma of the unclosed bladder.

The data of the 104 patients were obtained by detailed unstructured interviews during follow-up examination or by telephone-interviews with the patients or their relatives.

Mean follow-up from the time of the first operative intervention for bladder exstrophy or incontinent epispadias was 16.7 years (range 0.2–35). A total of 34 patients could be followed for more than 20 years.

RESULTS

Upper Urinary Tract and Continence

The upper urinary tract remained stable at the latest follow-up in 98% of those with a rectal reservoir, 95% with Mainz pouch I, 96% in colonic conduits and 100% in the 6 patients with a modified Young Dees or sling plasty. 94% with a rectal reservoir and 97% of those with a Mainz pouch I were continent day and night, whereas only 3 of 6 patients with a sling plasty (incontinent epispadias) or with primary bladder closure followed by a Young Dees procedure are continent (50%).

Education, Occupation, and Social Development

All children were raised within their families. Their educational achievements have been satisfactory for patients and their parents. Two children are younger than 4 years. Of the older children, 4 attend kindergarten, 41 elementary school and 7 secondary school. In most cases the teachers or schoolmates have not been informed of the anomaly. Four of the adults attend university, 5 are self-employed, 24 hold qualified jobs and 7 are employed as workers or drivers. 6 females are housewives who have completed vocational training, 3 of whom take care of their children. Only 3 patients are unemployed, 1 of whom has twice dropped out of job training programs. One male patient with multiple malformations and a lower IQ has been living and working in a therapeutic center since the age of 9. He underwent castration at the age of 4 in an outside hospital and was raised as a girl. At the age of 29, however, the patient decided to legally change his sexual identity and name but no further operation was undertaken.

Sexual Behaviour and Fertility

Females. Together with or after urinary diversion or primary bladder closure, reconstruction of the external genitalia was performed in 25 out of 37 girls at an average age of 12.2 years (2.5 - 27 years). Reconstructive surgery included approximation of the divergent mons pubis and bifid clitoris as well as reconstruction of the labia minora whenever possible. Vaginal cutback or vaginoplasty to enlarge the introitus vaginae was performed in 23 patients, in 13 instances combined with reconstruction of the external genitalia. Prolapse occurred several months later in 4 patients, 3 of whom subsequently underwent fixation of the uterus. A pessary was recommended by the gynaecologist to the remaining female. As a result of this experience we performed fixation of the uterus in the same session as vaginal cutback in 10 further cases. The average age at operation was 19 years (13–27 years).

In the adult group, 18 out of 19 patients are married or have a constant sexual partner. All engage in sexual intercourse, but 3 consider it unpleasant. After correction of the external genitalia and vaginal cutback, 16 of 17 women are satisfied with the cosmetic results.

Five women have delivered 7 children by Caesarean section. In three of these women antefixation of the uterus was done during the Mainz Pouch I procedure. 5 women presently desire children (Table 1).

Table 1. Fertility, pregnancy, delivery, and complications of pregnancy

Patient	Age at delivery	Urinary diversion/ sling plasty	Fixation of the uterus	Children	Complications	Delivery
1	18	Mainz pouch I	yes	1	no	sectio
2	29/32	Mainz pouch I	yes	2	no	sectio
3	33	Mainz pouch I	yes	1	no	sectio
4	26/27	Colon conduit	no	2	no	sectio
5	21	Sling plasty	no	1	mild upper tract dilatation and uterine prolapse	sectio

Hysterectomy, n = 2.
Presently desire children, n = 5.

Males. Whenever primary urinary diversion is performed at our institution, the bladder plate is almost entirely resected and the central portion near the verumontanum is closed to form a small reservoir.

Reconstruction of the external genitalia was performed in 32 out of 35 patients older than 18 years of age. One of them developed necrosis of the penis at an early age after primary bladder closure was performed at an outside hospital, however, he does not want a reconstruction. All other men achieve erection, and only 2 of the men are dissatisfied with the cosmetic result and have requested reoperation.

16 out of the 35 male adults are married or have a steady partner. Except for 8 younger patients, all engage in sexual intercourse. Penile deviation is present in 11 of the 32 men but is distressing in only 2.

A total of 27 patients reported that ejaculation was not normal, while 5 have no ejaculation at all. Spermiograms in 5 patients revealed astenospermia (70% immobile) in 1, azoospermia in 3 and a minimal amount of ejaculate (0.5 ml) with normal density in one. The remaining patients either refused a spermiogram or will undergo the procedure at the next follow-up examination. Only 10 men expressed the desire to have children. Epididymitis was found in 9 patients, which was recurrent in 6 and necessitated orchiectomy in 2 patients. Three patients had undergone vasectomy (2 bilaterally). No patient with reconstruction of the external genitalia can ejaculate normally or has fathered children. The wife of 1 patient became pregnant after heterologous insemination.

During the same period 5 patients without primary genital reconstruction were observed. All 5 men ejaculated normally and 2 have fathered 4 children. One of the two men with children died 2 years after cystectomy and Mainz Pouch I procedure due to metastases of a signet cell carcinoma of the unclosed bladder. Of the 5 patients, 2 underwent genital reconstruction as adults. Thereafter, one man was no longer able to ejaculate while the other had urethral strictures that required repeated internal urethrotomy.

As a result of our experience we created a small infrapubic opening in three boys leading to a small reservoir above the verumontanum at the penile base, thus offering the advantage of aspirating sperm easily. It can be closed at the patients wish after reaching puberty. None of these 3 patients has developed epididymitis.

DISCUSSION

As a result of modern operative techniques and highly improved postoperative medical care, children with bladder exstrophy today have a normal life expectancy.[6] Long-term follow-up data in patients with urinary diversion, especially ureterosigmoidostomy, document a normal education and social achievements in these individuals. The unemployment rate in the adults is very low as the literature and our own results confirm.[1,12]

Only few reports in the literature deal with reconstruction of the external female genitalia.[1,4] Reconstruction of the external genitalia with adaptation of the bifid clitoris and divergent mons pubis shows good cosmetic results, as our interviews confirm. Nevertheless, in some cases multiple operations for bladder reconstruction resulted in ugly scars on the abdominal wall, while in others the mons pubis remained split medially and divided by a huge hairless scar. For reconstruction of the lower abdominal wall some authors use tissue expanders.[13,14] In our experience, there was almost always enough skin to cover the abdominal wall defect. At our institution we attempt to perform the operation before the child enters kindergarten. Although the goal of this surgical intervention is "only" aesthetic, the first minimal vaginal cutback prevents retention of secretion and, after puberty, retention of menstrual blood. The portio is visible and the decision about an antefixation of the uterus and its timing can be made. At this age a uterine prolapse is unlikely. Furthermore, major psychological problems can be prevented, as the young girls no longer look and feel so different from the others after the operation. Due to shrinkage, a second definitive cutback is necessary before cohabitation to permit sexual intercourse.

Women with bladder exstrophy have a narrow vaginal introitus. This can be corrected by episiotomy or vaginoplasty.[1,2,4,15] In agreement with others we recommend the definitive introitusplasty before cohabitation.[2] At the time of definitive cutback antefixation of the uterus is recommended, as otherwise uterine prolapse of several degrees may occur. Correction of the mons pubis, excision of scars and creation of an umbilicus can be performed in the same session.

Uterine prolapse represents one of the main complications of pregnancy and occurs mostly after correction of the narrow introitus.[1,3,15] Operative procedures include the use of two lateral fascial strips,[16] a Goretex® sling,[1] and an ivalon sponge hystrosacropexy.[5]

Pregnancy in a woman with bladder exstrophy was first reported in 1724.[17] A survey among paediatric urologists as well as a review of the literature revealed 131 women who had delivered 164 children.[18] In a recently published series of 4 pregnancies after orthotopic continent urinary diversion (Indiana pouch) with perineal stoma, all 4 women developed uterine prolapse and urinary tract infections, 2 developed hydronephrosis and 2 had difficulties with clean intermittent catheterisation.[19] In contrast, our three patients with ileocaecal pouch and continent umbilical stoma did not show any complications.

In the male, reconstruction of the external genitalia and urethra is technically more complex and associated with a higher complication rate. Numerous operations have been reported. Gearhart and Jeffs perform genital reconstruction at the age of 2 to 3 years, after previous growth stimulation of the penis with testosterone cream.[20] On the other hand, Hanna and Williams recommend urethral reconstruction in primary urinary diversion patients after having reached the age of puberty.[11] At our institution we attempt reconstruction of the external genitalia between the ages of 4 and 6 years. At this time the penis has reached a sufficient size and the child has not entered school, thus avoiding unnecessary conflicts.

The cosmetic result in our patients after one or more reconstructions was satisfactory. Somewhat surprising was the fact that only 2 out of 11 patients with penile deviation found this disturbing during sexual intercourse. Similar reactions have been found by others.[6,8] These findings clearly show that even with a result considered unsatisfactory by the surgeon (continued penile deviation despite genital reconstruction) the patient can be thoroughly satisfied with organ function.

Iatrogenic infertility is the most serious late complication associated with reconstruction of the urogenital tract. For example, damage to the vas deferens and/or verumontanum during urinary diversion, cystectomy or during bladder neck reconstruction may result in lack of ejaculation.[1,11,21] Strictures of the neourethra may cause an increase in outlet resistance, resulting in retrograde ejaculation or, if closure is complete, in loss of ejaculation, which may lead to recurrent infections with ensuing damage to the epididymis, testes and even spermatozoa. Unfortunately, obstruction frequently remains unrecognised until puberty, unless the boy has had an epididymitis at an earlier age.

In our series we saw 9 out of 32 patients with epididymitis, 2 of whom required orchiectomy and 3 vasectomy. Corresponding data in the literature are missing. None of the men with genital reconstruction achieves normal ejaculation, 5 did not have any ejaculation at all, and none of the 32 fathered children. In contrast 2 out of 5 patients without reconstructive surgery have fathered 4 children. One of these patients, now 43 years old, whose bladder exstrophy remained untreated until the age of 22, underwent ureterosigmoidostomy and reconstruction of the external genitalia at that time. Before surgical intervention he had normal ejaculation, postoperatively he had none at all. The few reports in the literature show a similar tendency. Mesrobian et al. report about 9 out of 25 patients who also had no ejaculation; if the others had normal or prolonged ejaculation is not mentioned.[8] Of 12 patients reported by Audry et al., half ejaculated normally, while the other half lacked or had retrograde ejaculation.[10]

The few spermiograms obtained from our patients show a low motility, low quantity and abnormal morphology, thus severely limiting fertility. To our knowledge there has been only one report in the literature on the quality of spermiograms in bladder exstrophy patients.[11] While the spermiogram was normal in 4 of 8 patients with urinary diversions, this was true for only 1 out of 8 with primary bladder closure, and this patient had a high leukocyte count. The remainder showed oligospermia, azoospermia or no ejaculation at all.[1]

There are only few reports on the offspring of male patients with bladder exstrophy. Bennet reports on 3 of 32 men with ureterosigmoidostomy who fathered children, but data on genital reconstruction are lacking.[6] In the study of Woodhouse 6 out of 72 men have children, 5 of whom had undergone primary urinary diversion.[1] A similar tendency was found by other authors.[10] Our own results underline the restrictive effect of genital reconstruction on fertility as two of the male adults (without genital reconstruction!) have fathered children.

In patients with bladder neck reconstruction, retrograde ejaculation into the closed bladder may be responsible for impaired fertility.[21] The options for these patients are limited. In men with bladder neck reconstruction, spermatozoa may be retrieved from the urine, and in those with urinary diversion, sperm may be aspirated from the small reservoir above the verumontanum or in both groups from the epididymis, if it is undamaged.[2]

CONCLUSION

Patients with exstrophy or incontinent epispadias are extraordinarily motivated, intelligent, young people. Today modern techniques of urinary diversion ensure a normal life expectancy and full social integration if renal function is good. The side issues of these patients recede too often into the background. Our patients show impressively that the cosmetic and functional results after reconstruction of the external and, in females, the internal genitalia are satisfactory.

The first discussion with the parents in which the necessary operations are explained, should not only include the question of primary bladder closure and bladder neck repair versus urinary diversion, but also the reconstruction of the external and internal genitalia. The possibility of cohabitation and pregnancy in females and the risk of infertility in males must be mentioned explicitly.

REFERENCES

1. C. R. J. Woodhouse: Exstrophy and epispadias. In: Long-term paediatric urology. Edited by C. R. J. Woodhouse. Oxford, England: Blackwell Scientific Publications, pp. 127 - 150, 1991.
2. C. R. Blakeley and W. G. Mills: The obestric and gynaecological complications of bladder exstrophy and epispadias. Br J Obestet Gynaecol, **88**: 167–173, 1981.
3. K. A. Burbrige, T. W. Hensle, W. J. Chambers, R. Leb and K. F. Jeter: Pregnancy and sexual function in women with bladder exstrophy. Urology, **28**: 12, 1986.
4. R. J. Rhodes: Cosmetic vulval surgery following repair of ectopia vesicae. Proc. Roy. Soc. Med., **70**: 534–537, 1977.
5. J. Dewhurst, P. J. Toplis and J. H. Shepherd: Ivalon sponge hysterosacropexy for genital prolapse in patients with bladder exstrophy. Br. J. Obstet. Gynaecol., **87**: 67–69, 1980.
6. A. H. Bennett: Exstrophy of bladder treated by ureterosigmoidostomies: long-term evaluation. Urology, **2**: 165–168, 1973.
7. H. Lepor, E. Sharpio and R. D. Jeffs: Urethral reconstruction in boys with classical bladder exstrophy. J Urol, **131**: 512–515, 1984.
8. H. J. Mesrobian, P. P. 4. Kelalis and S. A. Kramer: Long-term followup of cosmetic appearance and genital function in boys with exstrophy: review of 53 patients. J Urol, **136**: 256–258, 1986.
9. H. G. Mesrobian, P. P. Kelalis and S. A. Kramer: Long-term followup of 103 patients with bladder exstrophy. Journal of Urology, **139**: 719–722, 1988.
10. G. Audry, C. Grapin, S. Loulidi, M. Gruner and J. Brueziere: Avenir genital des garcons atteints d'exstrophie vesicale ou d'epispadias avec incontinence. Ann Urol, **24**: 120–124, 1991.
11. M. K. Hanna and D. I. Williams: Genital function in males with vesical exstrophy and epispadias. Br J Urol, **44**: 169–174, 1972.
12. A. Zabbo and R. Kay: Ureterosigmoidostomy and bladder exstrophy: A long-term follow-up. Journal of Urology, **136**: 396–398, 1986.
13. J. F. Eid, P. Rosenberg, K. Rothaus, D. Mininberg, L. Hoffmann and E. D. Vaughan: Use of tissue expanders in final reconstruction of infra pubic midline scar, mons pubis and vulva after bladder exstrophy repair. Urology, **41**: 426–430, 1993.
14. F. Marconi, P. Messina, P. Pavanello and R. de Castro: Cosmetic reconstruction of the mons veneris and lower abdominal wall by skin expansion as the last stage of the surgical treatment of bladder exstrophy: a report of three cases. Plast Reconstr Surg, **91**: 551–555, 1993.
15. M. P. Krisiloff, P. J. Puchner, W. Tretter, M. T. MacFarlane and J. K. Lattimer: Pregnancy in women with bladder exstrophy. J Urol, **119**: 478–479, 1978.
16. E. W. Overstreet and J. F. Hinman: Some Gynaecologic aspects of bladder exstrophy. West. J. Surg. Obestet. Gynaecol., **64**: 131–137, 1956.
17. C. Clemetson: Ectopia vesicae and split pelvis. Br. J. Obestet. Gynaecol., **65**: 973- 981, 1958.
18. E. Sharpio, H. Lepor and R. D. Jeffs: The inheritance of the exstrophy-epispadias complex. J. Urol., **132**: 308–310, 1984.
19. W. A. Kennedy, H. T. W., E. A. Reiley, H. E. Fox and T. Haus: Pregnancy after orthotopic continent urinary diversion. Surg. Gynecol. Obstet., **177**: 405, 1993.
20. J. P. Gearhart and R. D. Jeffs: Exstrophy of the bladder, epispadias and other bladder anomalies. In: Campbell's Urology. Edited by P. C. Walsh, A. B. Retik, T. A. Stamey and E. D. Vaughan. New York: Saunders, 2, pp. 1772–1821, 1992.
21. M. D. Bastuba, M. M. Alper and R. D. Oates: Fertility and the use of assisted reproductive techniques in the adult male exstrophy/epispadias patient. Fertil. Steril., **60**: 733–740, 1993.
22. G. Palermo, H. Jorisk, M.-P. Derde, M. Camus, P. Devroey and A. van Steirteghem: Sperm characteristics and outcome of human assisted fertilization by subzonal insemination and intracytoplasmatic sperm injection. Fertil. Steril., **59**: 826–830, 1993.

29

SEMEN ANALYSIS AND TRANSRECTAL ULTRASOUND FINDINGS IN MEN WITH BLADDER EXSTROPHY

Terry W. Hensle*, Erik T. Goluboff, and Harry Fisch

Department of Urology
Babies and Children's Hospital
Columbia-Presbyterian Medical Center, New York

ABSTRACT

Although it has been reported that women with bladder exstrophy have relatively normal potential for fertility, little has been reported about such potential in men with this disorder. In 1991, Hamper et al. evaluated 7 men with bladder exstrophy with transrectal ultrasound (TRUS) and found that, although 6 had prostates and seminal vesicles, only 2 showed a normal appearance to these structures. In 1979, Lattimer et al. reported on semen analyses from 9 men with bladder exstrophy and found that 7 were oligo- or azoospermic. Since no study to date has simultaneously examined semen analyses, TRUS findings, and hormonal parameters in men with bladder exstrophy, we describe this data from 4 men with this disorder.

Semen analyses were significant for a pH greater than 8.5 and a low semen volume in all patients. Sperm density and motility ranged from aspermic to normal. TRUS findings were significant for prostates that ranged from cystic to small to normal. Most patients had abnormal distal insertion of the seminal vesicles/ejaculatory ducts into the prostate. FSH concentrations were normal in 2 of 4 patients while testosterone levels were all normal.

In conclusion, fertility potential in male exstrophy patients appears variable. The major finding here is the abnormal distal insertion of the seminal vesicles/ejaculatory ducts into the prostate associated with an abnormally high seminal pH. Three possible explanations for this are: 1) a developmental defect in this disorder may involve hindered growth of prostatic tissue, accounting for the elevated seminal pH, low semen volume, and abnormal seminal vesicle-prostatic insertion; 2) alterations in anatomy could be secondary

---

* Contact author: Terry W. Hensle, MD, 3959 Broadway, Babies Hospital, Columbia-Presbyterian Medical Center, New York, NY 10032, (212) 305-8510

The Exstrophy–Epispadias Complex, edited by Gearhart and Mathews.
Kluwer Academic / Plenum Publishers, New York, 1999.

surgical procedures undergone by these patients; and 3) functional changes in the prostate could result from multiple episodes of epididymitis.

INTRODUCTION

The bladder exstrophy-epispadias complex is a rare congenital abnormality which is caused by an overdevelopment of the cloacal membrane preventing an ingrowth of mesenchyme and normal development of the lower abdominal wall musculature. This syndrome occurs in 1 in 10,000 to 1 in 50,000 live births and is more common in males than females. The history and evolution of therapy for classic bladder exstrophy can be traced from the year 2000 B.C.[1] The surgical approach to this condition initially focused on early urinary diversion to preserve renal function. Since Young's report of a successful bladder closure in 1942[2], significant advances have been achieved in the areas of staged reconstruction, appearance of the external genitalia, and preservation of continence and renal function. With improvements in surgical technique, as well as perioperative care, excellent long-term survival can now be expected.[3] The extended survival of these children has lead to reports of fertility and sexual function in both male and female exstrophy patients.[3–6]

Fertility status in men with bladder exstrophy is variable. Semen analyses have ranged from azoospermia, to oligospermia, to low normal sperm densities and proven fertility.[4] Hamper et al. examined 7 patients with transrectal ultrasound (TRUS) and found that although all had prostates and seminal vesicles, most had an unusual appearance to these structures.[7] No study to date has simultaneously examined semen analysis, TRUS findings, and hormonal parameters in men with bladder exstrophy; the goal of our study was to describe this data in 4 men with this disorder.

MATERIALS AND METHODS

From July 1993 to March 1994, three patients with bladder exstrophy presented for evaluation of infertility; one patient with exstrophy with proven fertility was seen in the office for routine follow-up. All patients underwent detailed history and physical examinations with particular emphasis on sexual function, urinary continence, renal function, and past surgical history. Each patient underwent TRUS and gave samples of semen and blood.

Semen samples were produced by masturbation into wide-mouthed collection vessels and were processed immediately. Semen analyses were performed as per World Health Organization guidelines.[8] TRUS was performed using an Ausonics AU 530 ultrasound machine (International Ultrasound, Bayonne, NJ) with a 7.5 MHz probe. Careful attention was made to the size of the prostate and the spatial relationships of the seminal vesicles, prostate, and bladder to one another. Seminal vesicle dilation was defined as width greater than 15 mm, as per Jarow.[9] Hormonal testing was performed on venous blood for testosterone and follicle-stimulating hormone (FSH) by a commercial laboratory (Bioreference Laboratories, Elmwood Park, NJ).

RESULTS

Demographic, semen analysis, hormonal, and TRUS data are presented in Table 1. One man had fathered children previously while the other three had not. Sperm density

Table 1. Characteristics of men with bladder exstrophy*

Pt#	Age	Fert	Atrophy	FSH	Test	Vol	pH	Dens	Motil	Prostate	SVs
1	34	Y	Left	5	523	1.5	8.5	203	40%	13 cc	low
2	32	N	Right	12	535	0.5	8.5	1.5	20%	15 cc	low, dilated
3	26	N	N	16	314	0.3	8.5	1.5	30%	36 cc	normal
4	24	N	Right	30	335	0.3	8.5	0	0	cystic	low, dilated

*Age in years, Fert=known fertility, Atrophy=testicular atrophy, FSH=follicle stimulating hormone concentration in mIU/ml, Test= testosterone level in ng/dl, Vol=semen volume in cc, Dens=sperm density in 10^6/ml, pH=semen pH, Motil=sperm motility in %, prostate, SVs=TRUS appearance of prostate and seminal vesicles, low SVs= abnormal distal insertion of the seminal vesicles/ejaculatory ducts into the prostate.

was variable and ranged from 0 to 203 × 10^6/cc. When sperm were present, motility ranged from 20 to 40%. FSH levels were noted to be elevated in 2 patients while testosterone levels were normal in all. One consistent finding was that semen volume was low and seminal pH was high. TRUS findings revealed the presence of prostates and seminal vesicles in all 4 patients. Only one patient, patient 3, had a normal appearing prostate and seminal vesicles. The remaining 3 had abnormally low insertions of the ejaculatory duct into the prostate and 2 had dilated seminal vesicles bilaterally (Figure 1).

Patient 1 was 34 years old. At 6 months of age, he underwent bladder closure and bilateral osteotomies. His left testis was cryptorchid, was surgically placed in the scrotum, and is currently atrophic. He has had multiple procedures for epispadias. He has worn a condom catheter for the last 17 years and his bladder capacity is approximately 100 cc. He reports successful intercourse once per week for which he takes Bactrim for 2 days after-

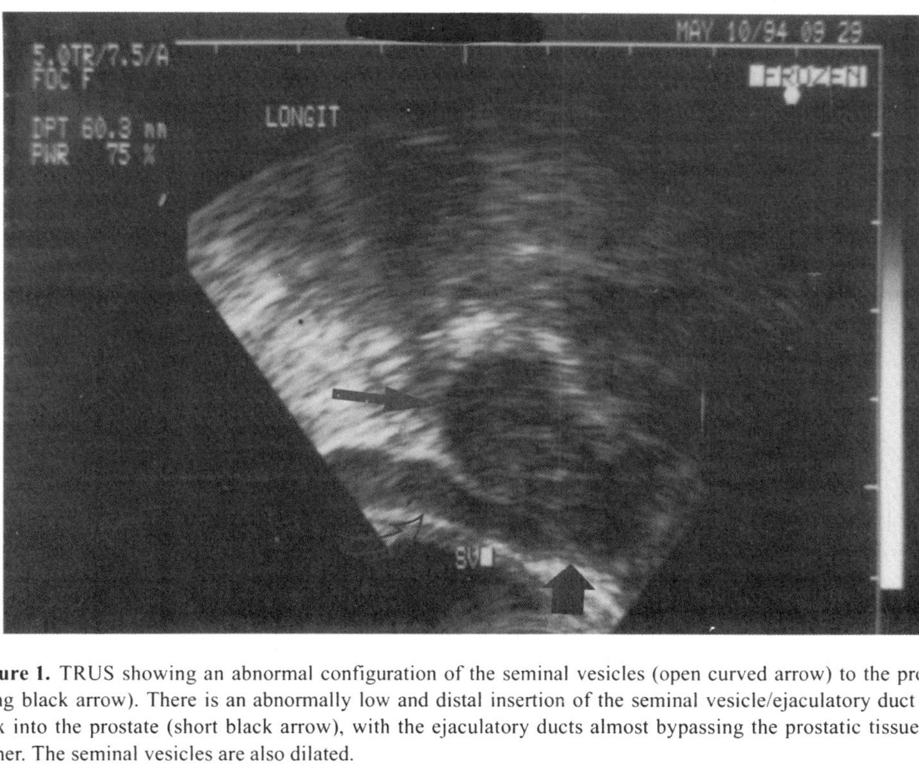

Figure 1. TRUS showing an abnormal configuration of the seminal vesicles (open curved arrow) to the prostate (long black arrow). There is an abnormally low and distal insertion of the seminal vesicle/ejaculatory duct complex into the prostate (short black arrow), with the ejaculatory ducts almost bypassing the prostatic tissue altogether. The seminal vesicles are also dilated.

wards to prevent epididymitis. He has fathered 2 children. Patient 2 was 32 years old. He underwent bladder closure at birth and ileal conduit diversion at age 3. His right testis is atrophic. He has intercourse 2–3 times per week. Patient 3 was 26 years old. He underwent bladder closure at birth and epispadias repair at age 3. His testicles were both normal. He has had multiple episodes of epididymitis. He voids per urethra but is incontinent and wears 2–3 pads per day. He has intercourse once weekly. Patient 4 was 24 years old. He underwent bladder closure at birth and has had multiple surgeries for epispadias. He was incontinent to a diaper until 2 years ago when he underwent a Mitrofanoff continent gastroilealcystoplasty to the umbilicus with the bladder neck left open. He currently voids per urethra and irrigates through the stoma every day. He is continent and post-void residuals are low. His right testis is atrophic. He has intercourse once weekly.

DISCUSSION

As a result of improvements in surgical technique and perioperative care, most infants born with bladder exstrophy reach adulthood. This improved survival has lead to an increasing interest in the reproductive capacity of both men and women with bladder exstrophy. Traditionally, the high rate of infertility in male exstrophy patients has been attributed to retrograde ejaculation or obstruction created at the time of bladder neck closure or other reconstructive procedures. As early as 1972, Hanna and Williams reported that male exstrophy patients who had undergone primary bladder reconstruction had lower sperm counts than patients who had undergone urinary diversion. Presumably, those who underwent diversion were less likely to have altered prostate-ejaculatory duct anatomy because there was no surgical manipulation in this area.[10] In our patients, only one was diverted and he is oligospermic (patient 2), while 2 others (patients 1 and 3) have only had single primary bladder closures and are normo- and oligospermic, respectively. One patient (4) who had extensive bladder reconstruction is azoospermic. Perhaps the fact that patients 1 and 3 had only minimal surgical manipulation of the bladder has enabled them to maintain more normal ejaculatory anatomy (patient 3) and function (patient 1).

Lattimer et al. reported on semen analyses in 9 men with exstrophy and found that 7 were oligo- or azoospermic. Etiologies for the poor semen analyses suggested by these authors included the high rates of cryptorchidism, episodes of epididymitis, and use of nitrofurantoin in men with bladder exstrophy. They suggested early use of broad-spectrum antibiotics and the avoidance of the use of nitrofurantoin.[4] In our patients, all have had multiple infections, either pyelonephritis or epididymitis, and 3 of 4 have had multiple stones in the past. None reported using nitrofurantoin. Patient 1, the only fertile one, had a history of epididymitis after intercourse, and was the only one to use prophylactic antibiotics.

The first detailed report of the anatomy of the prostate gland and seminal vesicles in men with exstrophy came in 1991 when Hamper et al. applied TRUS to 7 patients and found the prostate to be placed more anteriorly in the pelvis with a variable size and echotexture.[7] The seminal vesicles were large, dilated, and often cystic in appearance, similar to those seen in our study. They attributed these findings to the underlying embryologic nature of the condition as well as to post-surgical changes secondary to staged reconstruction. Gearhart et al. went on to analyze the configuration of the prostate in 13 bladder exstrophy patients using magnetic resonance imaging. They found all 13 patients to have normal prostate volumes, but none of the prostates extended circumferentially around the urethra.[11]

Our study is limited in several ways. First, there are relatively few patients studied. These patients, however, were the only ones with this data available. Our findings are necessarily preliminary and represent observations that need to be confirmed. Second, from an infertility standpoint, the patients were not evaluated exhaustively. For example, post-ejaculate urine samples were not examined for spermatozoa. As well, semen samples were not tested for fructose. In patient 1, these tests did not seem necessary since he had a normal sperm density and proven fertility. In patient 2, a post-ejaculate urinalysis would not have yielded any useful information since he had had an ileal conduit diversion. In patients 3 and 4, these tests were simply not performed. Third, multiple semen analyses were not obtained in these patients to confirm our findings. These omissions are largely on the basis of patient compliance.

These caveats aside, our study looked at both semen analyses and TRUS findings in men with bladder exstrophy. We can make 2 novel observations: in 3 of our patients there was an abnormally distal insertion of the seminal vesicle/ejaculatory duct complexes into the prostate and in all 4 patients there was an abnormally high seminal fluid pH. These concurrence of these two findings can be explained in several ways: 1) a developmental defect in this disorder may involve hindered growth of prostatic tissue, accounting for the elevated seminal pH, low semen volume, and abnormal seminal vesicle-prostatic insertion; 2) alterations in anatomy could be secondary surgical procedures undergone by these patients; and 3) functional changes in the prostate could result from multiple episodes of epididymitis.

REFERENCES

1. Gearhart JP and Jeffs RD: Exstrophy of the bladder, epispadias, and other bladder anomalies. In; Campbell's Urology. Edited by PC Walsh, AB Retik, TA Stamey and ED Vaughan. Philadelphia: WB Saunders, vol. 2, pp. 1772–1821, 1992.
2. Young HH: Exstrophy of the bladder: The first case in which a normal bladder and urinary control have been obtained by plastic operation. Surg Gynecol Obstet 74: 729, 1942.
3. Woodhouse CRJ, Ransely PG and Williams DI: The patient with exstrophy in adult life. Br J Urol 55: 632, 1983.
4. Lattimer JK, MacFarlane MT and Puchner PJ: Male exstrophy patients: A preliminary report on the reproductive capability. Trans Am Assoc Genitourinary Surgeons 70: 42, 1979.
5. Burbige KA, Hensle TW, Chambers WJ, Leb R and Jeter KF: Pregnancy and sexual function in women with bladder exstrophy. Urology 28: 12, 1986.
6. Lattimer JK, Beck L and Yeaw S: Long-term followup after exstrophy closure: Late improvement and good quality of life. J Urol 119: 664, 1978.
7. Hamper UM, Gearhart JP, Danhert WF, Smith S and Jeffs RD: Bladder exstrophy-epispadias complex: Prostatic evaluation by transrectal ultrasonography. Prostate 19: 133, 1991.
8. World Health Organization. Laboratory manual for the examination of human semen and semen-cervical mucous interaction. 3rd ed. Cambridge: The Press Syndicate of the University of Cambridge, 1991.
9. Jarow JP: The role of ultrasonography in the evaluation of the infertile male. Semin Urol 12: 274, 1994.
10. Hanna MK and Williams DJ: Genital function in males with vesical exstrophy and epispadias. Br J Urol 44: 1969, 1972.
11. Gearhart JP, Yang A, Leonard MP, Jeffs RD and Zerhouni EA: Prostate size and configuration in adults with bladder exstrophy. J Urol 149: 308, 1993.

30

ANATOMY OF CLOACAL EXSTROPHY

The Surgical Implications

Douglas A. Husmann*and David R. Vandersteen

The Mayo Clinic
Rochester, Minnesota

CLOACAL EXSTROPHY: INCIDENCE AND HISTORICAL PERSPECTIVES

Cloacal exstrophy with an estimated incidence of 1 in 200–400,000, is the rarest and most severe form of the exstrophy epispadias complex.[1,2] Prior to 1960 the multiple systemic defects found in association with this anomaly inevitably lead to death of the infant.[3,4] Refinements in surgical technology, neonatal care, hyperalimentation and antibiotic therapy currently allow us to obtain survival rates in the range of 85–90%.[1,5] To appropriately plan for the surgical reconstruction of these patients, the pediatric urologist, surgeon and radiologist managing these individuals must beware of the multiple anomalies that coexist with this entity. The purpose of this paper is to outline the associated congenital defects and review their impact on the surgical treatment and management of this patient population.

CLOACAL EXSTROPHY: ANATOMIC FINDINGS

Cloacal exstrophy is characterized by the following findings: pubic diastasis, a superiorly placed omphalocele, an exstrophic foreshortened hindgut or cecum located between two exstrophic hemi bladders, an imperforate anus, genital abnormalities and spinal defects. This rather classic configuration of findings have lead to the naming of this syndrome as the OEIS complex (Omphalocele, Exstrophy, Imperforate Anus and Spinal Defects).

* Contact author: Dr. Douglas A Husmann, Department of Urology, E 17 B, The Mayo Clinic, Rochester, MN 55905

The Exstrophy–Epispadias Complex, edited by Gearhart and Mathews.
Kluwer Academic / Plenum Publishers, New York, 1999.

To determine the frequency of coexisting congenital abnormalities in children with cloacal exstrophy we reviewed 194 cases that were outlined in the literature from 1964 through 1997.[1–20] This data was then combined with information gathered from a retrospective chart review of 49 children with cloacal exstrophy seen at either: A) The Hospital for Sick Children Toronto, Ontario from 1962–1987. B) Dallas Children's Hospital, Dallas, Texas from 1962–1992. C) The Mayo Clinic, Rochester MN from 1962–1997 (Table 1).

RADIOLOGIC EVALUATION

Radiologic evaluations in children with cloacal exstrophy should initially consist of plain films of the chest and abdomen along with an abdominal ultrasound. Additional films of the skull and extremities may be warranted if physical examination reveals coexisting abnormalities. Due to the high frequency of spinal anomalies either an ultrasound or MRI evaluation of the spine is mandatory. In addition, we have frequently found that a MRI of the pelvis is helpful in delineating the anatomic abnormalities frequently found within this region. This is especially true in the female pelvis where gynecological anomalies abound.[12]

SURGICAL APPROACHES

The surgical management of cloacal exstrophy is divided into two phases: A) the initial closure of the defect and B) the subsequent multiple reconstructive procedures necessary to achieve urinary continence, and functional genitalia. Due to the high degree of variability that exists between each patient the surgical management must be individualized.[21]

The surgeons involved in the initial closure of the defect must decide if a simultaneous or delayed closure is to be performed Simultaneous closure consists of excision of the omphalocele, formation of the terminal enteric stoma, functional closure of the bladder halves and either gender reassignment or reapproximation of the corporal bodies. A delayed closure consists of excision of the omphalocele, formation of a terminal enteric stoma and reapproximation of the bladder halves. The bladder is left exstrophic and usually closed between 1–6 months of age.[5,20]

In our review, we found that 45% (42/93) of the patients underwent a delayed closure. The decision to perform a delayed closure was invariably due one of three reasons, a large omphalocele, 67% (28/42), cyanotic heart disease, 19% (8/42) or prematurity, 14% (6/42). Although 55% (51/93) of the patients underwent simultaneous closure of the cloacal defect, prosthetic materials were necessary to aid in closure of the abdominal wall in 13% of these individuals(7/51). [Note: In an attempt to pursue simultaneous closure in all of our patients, we have had to build a silastic silo to contain the intraperitoneal contents in three cases. All three developed Pseudomonas colonization of the prosthesis and eventually had a gram negative septic episode. Due to these experiences we would now choose to have patients undergo a delayed repair rather than perform a simultaneous closure using prosthetic materials].

MANAGEMENT OF THE HIND GUT: TERMINAL ILEOSTOMY VERSUS COLOSTOMY

The first major advance leading to increased survival of the patient with cloacal exstrophy was the recognition of the importance of the hindgut. This small segment of bowel

Table 1. Cloacal exstrophy anatomic findings

System evaluated	No. pts. with defect (%)	No. pt. evaluated
Genotype	3(3)	86
47 XX + 21	1(1)	
45 XO/46XX	1(1)	
47 XXX	1(1)	
Craniofacial	4(1.5)	243*
Ocular auriculo abnormalities	3(1)	
Frontonasal clefts	1(0.5)	
Pulmonary	5(9)	53
Atretic segmental bronchus	4(7)	
Diaphragmatic hernia	1(2)	
Cardiovascular	18(18)	99
VSD	6(6)	
Dextrocardia	3(3)	
PDA	3(3)	
Duplication of great vessels (aorta or IVC)	3(3)	
Hemangiomas of pelvis and lower limbs	3(3)	
Upper urinary tract	93(67)	139
Pelvic kidney	35(25)	
Unil. or bil. renal agenesis	18(13)	
Complete ureteral duplication (with or without ectopy)	18(13)	
Unil. multicystic dysplasia	10(7)	
UPJ obstruction	10(7)	
Single system ectopic ureter	5(4)	
Fusion abnormality	3(2)	
Gastrointestinal (excluding imperforate anus)	111(96)	116
Omphalocele	111(96)	
Inguinal hernia	42(36)	
Malrotation	17(15)	
Short small bowel	15(13)	
Duplication	12(10)	
Duodenal atresia	4(3)	
Meckel's diverticulum	2(2)	
Gynecological	102(100)	102
Bifid clitoris	72(70)	
Clitoral agenesis	17(17)	
Single clitoral body	13(13)	
Vaginal duplication	55(54)	
Vaginal agenesis	20(20)	
Vaginal duplication and nilateral vaginal agenesis	5(5)	
Uterine duplication	96(94)	
Uterine agenesis	1(1)	
Unilateral agenesis of salpinx and ovary	1(1)	
Male genitalia	99(100)	99
Bifid phallus	67(67)	
Penile agenesis or rudimentary penis	20(20)	
Single corporal body	12(12)	
Bifid scrotum	60(60)	
Absent scrotum	35(35)	
Single scrotum	4 (4)	
Cryptorchidism (uni or bilateral)	73(73)	
Vertebral (also see CNS)	186(93)	199
Spinal dysraphism	113(50)	
Sacral agenesis (partial or complete)	69(35)	
Hemivertebrae	15(8)	

(continued)

Table 1. (*Continued*)

System evaluated	No. pts. with defect (%)	No. pt. evaluated
Central nervous system (also see vertebral)	45/49(92)	49
Abnormal sacral sensation	45/49 (92)	
Myelodysplasia	34/49 (69)	
Tethered cord (symptomatic only)	16/43 (33)	
Hydrocephalus	1(2)	
Lower extremity	61(36)	170
Talipes equinovarus	32(19)	
Congenital dislocated hips	17(10)	
Phocomelia of one or both lower limbs	6(4)	
Severe deformity leg and foot	3(2)	
Calcaneous deformities	3(2)	

The data noted in the table is gathered by reviewing the patients in references 1, 5-19 (N=194) and adding the information obtained by chart reviews from 3 separate hospitals (N= 49): The Hospital for Sick Children, Toronto, Ontario, Children's Hopital, Dallas, TX and the Mayo Clinic, Rochester, MN. The table reflects a total of 243 patients. The information included within the table is based upon whether or not the authors specifically noted the presence or absence of the congenital abnormality. There is only one notable exception to this rule, marked by the *. In the literature reviews craniofacial abnormalities would have been stated within the published reports and therefore list all othe patients reviewed as the denominator.
In some categories the number of isolated deformities is greater than that for the whole system. This discrepancy is due to the fact that a patient may have had more than one of the abnormalities present.

averages 10 cm in length (range from 1–30 cm) and frequently appears rudimentary in structure.[21] In the remote past the appearance of this segment frequently led to its excision at the time of functional closure. In individuals closed prior to 1972, excision of the hindgut and formation of a terminal ileostomy resulted in a mortality rate of 68% (15/22). In contrast, preservation of the hindgut with formation of a terminal colostomy resulted in a mortality rate of 24% (6/25), $p < 0.01$.

In 1972 the use of hyperalimentation in the management of neonatal patients became a wide spread phenomena. The liberal use of hyperalimentation within the cloacal exstrophy population has had a significant impact on the prognosis of patients managed by terminal ileostomy. Currently the mortality between patients managed by terminal ileostomies and terminal colostomies are now identical, 5% (1/21) versus 4% (1/28) respectively.[5,19]

Although the current mortality rates between patients with an ileostomy and a colostomy are similar, a significant difference in the morbidity between the two surgical treatments remain. Specifically, when comparing patients with terminal ileostomies ($N = 21$) versus colostomy ($N = 28$), patients with an ileostomy require more frequent in patient hospital stays for short bowel syndrome (median number of in hospital days 39, range 0–265 versus a median number of hospital days of 10, range 0–20, $p < 0.01$) and they require the use of intravenous hyperalimentation for longer time periods (median number of days on intravenous hyperalimentation 60 days, range 14–240 versus median number of days 14, range, 0–67, $p < 0.01$). These differences predominate during the first three years of life, after that time the small bowel appears to adapt and no significant differences can be found between the two patient populations.[5,19]

SPINAL ABNORMALITIES

Vertebral and neurologic abnormalities are present in 92–93% of the patients with cloacal exstrophy.[10] Their impact on the patient's ability to obtain urinary and fecal continence, ambulation and in some circumstances sexual function are profound.[5,10,16,19,21]

It is important to note that the table records only the number of patients who have developed symptoms of a tethered spinal cord, 33% (16/43). If MRI evaluations are performed in all patients tethering of the spinal cord is found to exist in approximately 90% of the patients.[10] We currently do not believe that all of the cloacal exstrophy patients with a tethered cord on MRI evaluation require surgery. Rather, we request that the cord in these patients be untethered when the patient develops symptoms, that is, back pain or changes in lower extremity strength or sensation.[22] It remains to be seen whether or not more of these patients will develop problems with a tethered cord with prolonged follow-up.

NEUROLOGICAL DEFICITS: IMPACT ON MANAGEMENT OF THE IMPERFORATE ANUS

The long term management of the fecal stream in patients with cloacal exstrophy is difficult. Fecal continence is dependent upon the length of the hind gut, the presence of any coexisting neurological deficits and the adequacy of the pelvic musculature. In the majority of cases a pull through procedure is simply not feasible.[21] In our review we found a total of 24 patients that underwent a Peña pull through procedure. All of the patients had intact sacral sensation, good perineal musculature (on MRI evaluation) and an adequate length of colon, fecal continence is present in only 6 (25%). The remaining patients are requiring strict bowel management programs and daily rectal enemas to obtain some degree of cleanliness. In the future it may be possible that antigrade continence enemas (ACE) procedures will be able to be performed in some of these patients. At this time little data is available regarding the success of the ACE procedure within this specific patient population.[5,16,19,21]

NEUROLOGICAL DEFICITS: IMPACT ON STAGED BLADDER RECONSTRUCTION

To evaluate the adequacy of staged bladder reconstruction in the management of cloacal exstrophy we recently reviewed our patient population.[22] All of the patients within this study had been managed in a similar fashion that is, simultaneous functional closure within 48 hours of birth, and Young Dees Leadbetter bladder neck reconstruction at 3–5 years of age. If the bladder capacity was not greater than 60 cc at the time of bladder neck reconstruction a simultaneous bladder augmentation was performed. Within our patients' 57% (13/23) required an augmentation, the stomach was used in 5 (38%), ileum in 5 (38%) and hindgut in 3 (24%). Urinary continence was achieved in 22% (5/23). Only two of our continent patients are voiding. Three require intermittent catheterization to remain dry. When we performed a multivarient analysis of our data we found that the neurologic status of the patient was the most important determinant in the ability of the patient to become continent. In the 13 patients with impaired neurologic function only one (7%) was continent. In the 10 patients without a neurologic impairment four (40%) are continent, $p < 0.05$. Currently we believe that the nearly universal presence of neurologic abnormalities in patients with cloacal exstrophy makes the likelihood of obtaining a continent voiding patient extremely unlikely. Indeed, in our extensive review of the literature we could only find a total of 4 patients that were both continent and voiding following staged bladder construction.[9,20,22]

In addition to the congenital neurologic defects that exist, considerable risk of iatrogenic denervation of the bladder at the time of functional closure is present.[10,24] Detailed micro dissection of the pelvic nerves in a patient with cloacal exstrophy have revealed that the autonomic innervation of the bladder is unlike that of the normal infant. In the normal infant, the pelvic nerves arise from the pelvic plexus lateral to the rectum and run adjacent to the bladder's blood supply. In cloacal exstrophy the pelvic plexus is located on the anterior surface of the hindgut. The neural fibers innervating the bladder run laterally away from the medially located hindgut, traveling along its inferior and posterior aspects to reach each bladder half. It is important to point out that the neural supply to the bladder in cloacal exstrophy does not run adjacent to its blood supply. At the time of functional closure, dissection of the hind gut from the bladder halves places the neurological supply to the bladder at great risk. Indeed, it is quite possible to mobilize the hindgut, preserve the bladder's blood supply, but partially or complete denervate the bladder.[24]

IMPACTS OF ANATOMY ON THE MITROFANOFF PRINCIPLE

In the patient with cloacal exstrophy the Mitrofanoff principle must usually be used to obtain urinary continence. Unfortunately, in our experience the appendix has frequently been either excised at the time of closure or its lumen obliterated by scarring. The appendix is therefor rarely available for use as a catheterizable conduit. The next best alternative, the ureter, is not usable in approximately 45% of the patients. Inability to use the ureter is usually due to either inadequate length (pelvic kidney) or ureteral stenosis (ureter associated with renal dysplasia or agenesis). We frequently have had to rely on a catheterizable ileal tube as a alternative choice.[22,25,26]

MALE GENITAL ANOMALIES

Male genital abnormalities are routinely found within the cloacal exstrophy patient population. Insufficient phallic tissue due to either aphallia, a rudimentary penis or a single corporal body exists in 32% of the patients. When faced with these particular abnormalities we strongly recommend gonadectomy and gender reassignment. In the remaining 68% of the male patients the decision is not as straight forward. The penis is usually bifid and may be widely separated due to the pubic diastasis. Based on the recommendations of Tank and Lindenauer we have routinely recommended sexual conversion of these children.[27] In spite of these recommendations, not all of the parents have consented, in essence, forcing us to raise some of these children as males.[28] At present a total of 15 male cloacal exstrophy patients have completed puberty and are available for review. Nine (60%) have attempted vaginal intercourse but have been unsuccessful due to inadequate phallus size. Four (27%) are currently married and engaging in successful vaginal intercourse. Two patients (13%) are unable to obtain an erection, neither will consent to a full impotence evaluation. These findings have caused us to have some confusion regarding the routine use of gender reassignment. At present we approach the management of the male exstrophy patient with a bifid penis with great caution and confusion. This perplexity is in part due to the successful sexual adjustment of 27% of the male patients. The temptation to raise these children as males is however, tempered by the fact that 73% (11/15) of the males feel sexually inadequate. In addition, the physician should be aware that two of these male patients have either attempted or successfully completed suicide, presumably due to their feeling of sexual inadequacy.[18,28] Our inability to decide how to manage these patients would be greatly alleviated if long term results regarding the sexual function and psychologic adjustment of the 46 XY gender converted (i.e., phenotypic females) patients were available. Without this vital information the physician is forced to rely on his/her emperic judgment rather than fact.

FEMALE GENITAL ANOMALIES

Female genital anomalies especially uterine duplication is almost invariably present. Currently little if any data exists regarding the sexual function and psychological adjustment of the female patient with cloacal exstrophy. In our own personal experience several of these individuals appear to undergo episodes of depression during their pubescent years. It is our opinion that unless we truly know the psychological profile of the normal female with cloacal exstrophy, we will be unable to determine how to best manage the genital anomalies found in this congenital defect.

SKELETAL ANOMALIES

All patients with cloacal exstrophy will have pubic diastasis. In the past we performed iliac osteotomies to correct the diastasis, reapproximate the pelvic floor and hopefully improve our ability to obtain urinary continence. Since the majority of our patients require a continent catheterizable stoma and do not void spontaneously, we no longer believe that iliac osteotomies are necessary.[22] Indeed, we are somewhat please that we have abandoned their routine use. After we stopped performing the osteotomies we came to the realization that they were part of the problem we were having with the abdominal wall closure. Specifically, in cloacal exstrophy a large abdominal wall defect is present due to the omphalocele. In the absence of iliac osteotomies it is possible to compress most of the abdominal contents into the remaining abdominal and pelvic region and successfully close the abdominal wall. In contrast, when iliac osteotomies are performed a significant restriction in the available pelvic space occurs. Inadequate intraperitoneal and pelvic space frequently resulted in our inability to replace all of the intraperitoneal contents within the abdomen. This unfortunate set of circumstances would force us to use prosthetic material to close the abdominal wall. Abandonment of iliac osteotomies at the time of functional bladder closure has significantly reduced the need for using prosthetic material. The ability to close the abdominal wall without prosthetic material has decreased the operative morbidity with little if any impact in our ability to obtain urinary continence.

REFERENCES

1. Hurwitz, R.S., Manzoni, G.A.M., Ransley, P.G., Stephens, F.D.: Cloacal Exstophy: A report of 34 cases. J of Urol, 138: 1060–1064, 1987.
2. Gearhart, J.P.: Bladder and urachal abnormalities: The exstrophy-epispadias complex *in* Clinical Pediatric Urology, 3rd edition, editors, P.P. Kelalia, L.R. King, A.B. Belman, W.B. Saunders Co, Philadelphia, 1992, pp 608–611.
3. Swan, H., Christensen, S.P.: Exstrophy of the cloaca. Pediatrics, 12: 645–651, 1953.
4. Rickham, P.P.: Vesico-intestinal fissure. Arch. Dis. Child 35: 97–102, 1960.

5. Husmann, D.A., McLorie, G.A., Churchill, B.M., Ein, S.H.: Management of the hindgut in cloacal exstrophy: Terminal ileostomy versus colostomy. J of Ped Surg: 23: 1107–1113, 1988.
6. Manzoni, G.A., Ransley, P.G., Hurwitz, R.S.: Cloacal exstrophy and cloacal exstrophy varients: A proposed system of classification. J of Urol, 138: 1065–1068, 1987.
7. Lin, H.J., Ndiforchu, F., Patell, S: Exstophy of the cloacal in a 47 XXX child: Review of genitourinary malformations in triple XXX patients. Amer J of Med Genetics 45: 761–763, 1993.
8. Hendren, W. H.: Ileal nipple for continence in cloacal exstrophy. J of Urol 148: 372–379, 1992.
9. Diamond, D.A., Jeffs, R.D.: Cloacal Exstrophy: A 22 year experience. J of Urology, 133: 779–782, 1985.
10. McLaughlin, K.P., Rink, R.C., Kalsbeck, J.E., Keating, M.A., Adams, M.C., King, S.J., Luerssen: Cloacal Exstrophy: The neurological implications. J of Urology, 154; 782–784, 1995.
11. Haldar, A., Sharma, A.K., Phadke, S.R., Jain, A., Agarwal, S.S.: OEIS complex with crainofacial anomalies-Defect in blastogenesis? Amer. J of Med Genetics, 53: 21–23, 1994.
12. Meglin, A.J., Balotin, R.J., Jelinek, J.S., Fishman, E.K., Jeffs, R.D., Ghaed, V.: Cloacal exstrophy: Radiologic findings in 13 patients. AJR: 155: 1267–1272, 1990.
13. Greene, W.B., Dias, L.S., Lindseth, R.E., Torch, M.A.: Musculoskeletal problems in association with cloacal exstrophy. J of Bone and Joint 73:551–560, 1991.
14. Smith, N.M., Chambers, H.M., Furness, Haan, E.A.: The OEIS complex (omphalocele-exstrophy-imperforate anus-spinal defects) Recurrence in sibs. J Med Genetic 29: 730–732, 1992.
15. Stolar, C.J.H., Randolph. J.G., Flanigan, L.P.: Individualized management through a staged surgical approach. J of Ped Surg 25: 505–507, 1990.
16. Lund, D.P., Hendren, W.H.: Clocal Exstrophy: Experience with 20 cases. J of Ped Surg 28: 1360–1369, 1993.
17. Soper, R.T., Kilger, K.: Vesico-intestinal fissure. J of Urology 92: 490–501, 1964.
18. Howell, C., Caldamone, A., Snyder, H., Ziegler, M., Duckett J: Optimal management of cloacal exstrophy. J Ped Surg. 18: 365–369, 1983.
19. Davidoff, A.M., Hebra, A., Balmer, D, Templeton, J.M., Schnaufer, L.:Management of the gastrointestinal tract and nutrition inpatients with cloacal exstrophy. J of Pediatric Surgery, 31: 771–773, 1996.
20. Mitchell, M.E., Brito, C.G., Rink, R.C.: Cloacal exstrophy reconstructioon for urinary contience. J of Urology 144: 554–558, 1990.
21. Hurwitz, R., Manzoni, G.M.: Cloacal Exstrophy *in* Pediatric Urology 3rd edition, edited by B. O'Donnell and S.A. Koff, Butterworth Heinemann,Oxford. 1997, pp 514–525.
22. Husmann, D.A., Vandersteen, D.R., McLorie, G.M., Churchill, B.M.: Cloacal exstrophy: Evaluation of urinary continence following stage bladder reconstruction. J of Urology 157: 972, 1997.
23. Husmann, D.A.: Occult Spinal Dysraphism: The tethered cord and the urologist. *in* AUA Update Series, editor T.P. Ball, American Urological Association, Houston, 1997, Volume 14, pp78–83.
24. Schlegel, Gearhart, J.P.: Neuroanatomy of the pelvis in an infant with cloacal exstrophy: A detailed microdissection with histology. J of Urology 141: 583–585, 1989.
25. Duckett, J.W., Snyder, H.M.: Continent urinary diversion: Variations on the Mitrofanoff priciple. J of Urology 135:58–62, 1986.
26. Monti, P.R., Lara, R.C., Dutra, M.A., De Carvalho, J.R.: New techniques for constructionof efferent conduits based on the Mitrofanoff principle. Urology 49: 112–115, 1997.
27. Tank, E.S., Lindenauer, S.M.: Principles of management of exstrophy of the cloaca. Amer J of Surg 119: 95–100, 1970.
28. Husmann, D.A., Mclorie, G.A., Churchill, B.M.: Phallic reconstruction in cloacal exstrophy. J of Urology 142: 563–564, 1989.

31

CURRENT UROLOGIC MANAGEMENT OF CLOACAL EXSTROPHY

Experience with 11 Patients

Edwin A. Smith, John R. Woodard, Bruce H. Broecker, Rafael Gosalbez, Jr., and Richard R. Ricketts

Department of Surgery, Emory University School of Medicine
Egleston Children's Hospital
Atlanta, Georgia

ABSTRACT

Purpose

Since 1980 the authors have treated 12 infants with cloacal exstrophy (10 classical and 2 variants). Eleven patients had repair, and are all surviving. The initial phases of management that led to improved survival have previously been reported. Quality of life is now a major focus for the cloacal exstrophy patient. During the past 10 years, nine of the 11 patients had lower urinary tract reconstructive procedures. This review evaluates experience with reconstructive efforts to achieve bowel and bladder control and to improve the quality of life in this complex group of patients.

Methods

Through review of patient charts and by patient interviews, data were collected to evaluate the ability to provide urinary and bowel control. A continence score was applied to provide a measure of success: voluntary control, 3; control with an enema program or intermittent catheterization, 2; incontinence with a well-functioning stoma, 1; and incontinence without a stoma, 0. The best continence score is 6 (genitourinary and gastrointestinal). Surgical complications, urodynamic and metabolic sequelae of continent urinary diversion were reviewed.

The Exstrophy–Epispadias Complex, edited by Gearhart and Mathews.
Kluwer Academic / Plenum Publishers, New York, 1999.

Results

At the time of the authors' previous report, eight of 11 patients had a continence score of 2 or less. Currently, eight of 11 patients have a score of 3 or better (five with enteric stoma and continent urinary diversion, two with enema program and continent urinary diversion, and one with enema program and continent bladder). Urinary diversion procedures have included two gastric augmentations and five gastric reservoirs, two of which have a required subsequent bowel augmentation. Gastric augmentations carry a definite risk of metabolic problems with three of our patients demonstrating significant episodes of metabolic alkalosis. In addition, results of urodynamic monitoring suggests that gastric reservoirs may be less compliant than reservoirs formed using other bowel segments.

Conclusions

Modern principles of continent urinary diversion have been successfully applied to the cloacal exstrophy patient further improving their quality of life. Use of gastric flaps with preservation of intestinal length has been central to urologic reconstructive efforts. Use of stomach alone for formation of urinary reservoirs may produce suboptimal compliance, and composite ileogastric construction should be considered if the gastric flap is of marginal size.

Considerable progress has been made in treating the patient with cloacal exstrophy since the first report of a child surviving surgical correction was made in 1960.[1] The foundation of this success has rested on advances in neonatal medicine, intensive nutritional support, and use of staged reconstructive efforts based on a greater understanding of the anomaly.[2–4] With survival rates approaching 90%,[5] reports describing treatment of the cloacal exstrophy patient have now shifted from a focus on improving survival to improving quality of life. We previously reported our results with a series of 11 patients who had completed the initial phases of reconstruction with definitive treatment of the gastrointestinal tract.[6] The expectation at that time was to reconstruct, when appropriate, the lower urinary tract to achieve continence. Although functional bladder closure may be achieved in the cloacal exstrophy population,[7] the majority of patients will require bladder augmentation to improve capacity and compliance and construction of a catheterization continence mechanism.[8] Furthermore, in those patients in whom the bladder is too rudimentary to be of reconstructive value or in whom the bladder is committed to vaginal reconstruction, a continent urinary reservoir must be constructed. The stomach has been described as the ideal graft source in these patients because they have limited enteric length. Augmentation gastrocystoplasty and gastric reservoir construction have provided a method of achieving urinary continence while preserving the intestinal length that is important to fluid and electrolyte balance and to nutrition.[9] We have reevaluated our patient series after lower urinary tract reconstruction.

MATERIALS AND METHODS

The study group represents those patients with cloacal exstrophy who presented to Egleston Children's Hospital at Emory University from 1980 to 1987. These same patients were the subject of a previous report that detailed our results with a staged reconstructive approach to cloacal exstrophy.[6] They have now reached a median age of 9.9 years (range 7.5 to 15 years). Included are six genetic girls and six genetic boys, five of whom had female gender reassignment. Ten of these patients showed features of classic cloacal exstrophy with an omphalocele, exstrophied bladder halves separated by a lateral intestinal fissure, and imperforate anus. Two patients showed variant features including one boy with an omphalocele, an intact but exstrophied bladder, a colovesical fistula, imperforate anus, and intact phallus with a normal phallic urethra. One girl had abdominal wall features of prune-belly syndrome, omphalocele, an intact bladder opening onto the abdominal wall through a cloaca but with an absent urethra, hydrometrocolpos, and imperforate anus.

Associated urinary tract anomalies included renal hypoplasia (n = 2), renal agenesis (n = 1), pelvic kidney (n = 3), ureteropelvic junction obstruction (n = 1), ureteral duplication (n = 1), and unilateral hydroureteronephrosis (n = 1). Bilateral hydronephrosis was present in the patient with hydrometrocolpos. The phallus in the boys was intact in one, rudimentary in one, and bifid in four.

Of the 12 patients, 11 underwent reconstruction and one patient died as a neonate from severe pulmonary hypoplasia before any reconstructive efforts. Neonatal treatment specifically with regard to the exstrophied bladder segments included: end ileostomy with the vesicoenteric plate left undisturbed (n = 3); separation of the bladder segments from the lateral enteric fissure with approximation of the bladder halves (n = 4); and separation of the bladder segments from the lateral enteric fissure approximation of the bladder halves, and primary bladder closure (n = 3). One patient had an intact bladder as a closed variant. Secondary bladder closure was performed in seven patients with concomitant iliac osteotomy at a mean age of 23 months.

In an effort to achieve urinary continence seven of the 11 patients have had either bladder augmentation or construction of a continent urinary reservoir. The remaining four patients include the variant with an intact urethra that voided volitionally; the variant with a closed bladder but absent urethra that has been treated with an appendiceal Mitrofanoff catheterization stoma; one patient with rudimentary bladder that was given an ileal conduit; and a fourth patient with mental retardation who remains incontinent with epispadias after closure of the bladder. In all reconstructed cases the stomach was the gastrointestinal segment selected for inclusion into the urinary tract. A purely gastric reservoir was constructed and positioned orthotopically within the pelvis in five patients, and two patients had closure of the bladder neck and an augmentation gastrocystoplasty. Two patients who underwent reconstruction with gastric reservoirs subsequently had ileal augmentations producing composite reservoirs. A catheterization stoma was developed from a distal ureteral segment in five patients with the stoma situated on the abdominal wall in three patients and advanced to the perineum in two patients. Tapered ileum was used in a Mitrofanoff fashion to create a continent catheterization stoma in two patients.

We reviewed our results with respect to the stability of the upper urinary tracts based on renal function and radiographic appearance, metabolic consequences, urodynamic features, and need for surgical revision. Patient interviews and clinic and hospital chart reviews (by author EAS) served as data sources. Radiographic assessment included a postoperative cystogram or reservoirgram before removal of the suprapubic catheter and initiation of intermittent catheterization. A diethylenetriamine pentaacetic acid (DTPA) scan assessed postoperative drainage of the upper tracts. Renal sonography was performed every 6 months for 1 year postoperatively and then was performed yearly. Video urodynamic manometry was performed with saline infusion at a rate of 20 mL/min via a 7F urodynamic catheter inserted through the catheterizable stoma. The impact on quality of life was assessed by assigning a continence score reflective of the status of both the gastrointestinal tract as well as the genitourinary tract. Points were assigned as follows: voluntary control, 3; control with an enema program or intermittent catheterization, 2; incontinence with a well-functioning stoma, 1; incontinence without a stoma, 0. Summation of the gastrointestinal and genitourinary points yields a continence score ranging from 0 (worst) to 6 (best) with rigid adherence to the scoring criteria.

RESULTS

Since our initial description, seven of the 11 patients had continent reconstruction of the urinary tract with use of gastric flaps. The mean age at which continent urinary reconstruction was undertaken was 6 years (range 3 to 8 years) and the mean postreconstruction follow-up interval is now 3.5 years (range 1.5 to 5 years). Two variant patients displayed features that were particularly favorable for reconstruction without augmentation, and two patients have not undergone reconstruction for urinary continence. Neither of the unreconstructed patients has sufficient social support and self reliance to offer continent diversion.

Surgical Complications

Early postoperative complications occurred in one patient with a gastric reservoir with incontinence of the tapered ileal catheterization stoma and development of a small bowel obstruction requiring exploratory laparotomy. After small bowel resection an enterocutaneous fistula developed. The fistula was subsequently closed and a hemi-Kock augmentation of the reservoir was performed but failed to provide a continent stoma. Two additional patients with gastric reservoirs had delayed reoperation. One developed a spontaneous reservoir perforation occurring at 6 weeks and again at 6 months postoperatively. Ileal augmentation was performed at the time of repair of the second perforation. The other patient developed a ureterogastric anastomotic stricture, which was detected and successfully repaired at 9 months after urinary reconstruction.

Upper Tract Stability

Ureteral reimplantation into the gastric reservoir was successful in all patients; none developed reflux. Upper tract stability has been noted in six of the nine patients with urinary continence. One patient with a gastric reservoir and one patient with an augmentation gastrocystoplasty have demonstrated a moderate increase in hydronephrosis bilaterally. The hydronephrosis appears to be related to low reservoir compliance and is currently being managed with more frequent catheterization. The creatinine remains normal for age in all patients (mean creatinine 0.6 mg/dL).

Metabolic Complications

Three patients, two with gastric reservoirs and one with an augmentation gastrocystoplasty, have required hospitalization for treatment of dehydration with hyponatremia, hypochloremia, and hypokalemia. These metabolic imbalances have, in every case, occurred during viral gastroenteritis, and no sustained metabolic derangements have occurred. The hematuria-dysuria syndrome has not occurred in this group of patients, and chronic antacid blockade has not been necessary.

Urodynamics

A urodynamic evaluation was performed 6 months after reconstruction in six of the seven patients; three have gastric reservoirs, two have gastric augmentations, and one has

Table 1. Urodynamic data and upper tract status after augmentation gastrocystoplasty or gastric reservoir construction

				Urodynamic data		
Patient	Current age (yr/mo)	Urinary diversion	(% Expected)	*ML capacity (mL/cm H_2O)+	Compliance	Urodynamic results (renal ultrasound)
KB	10/8	Gastrocystoplasty with ureteral Mitrofanoff	280(77)	22		No hydronephrosis/Cr, 0.5mg/dL
TW	12/8	Gastrocystoplasty with ureteral Mitrofanoff	200(48)	9		Moderate hydronephrosis/Cr, 0.9mg/dL
MP	8/9	Gastric reservoir with ureteral Mitrofanoff	400(133)		20	No hydronephrosis/Cr, 0.5mg/dL
MS	11/3	Gastric reservoir with ureteral Mitrofanoff	300(77)	10		Moderate hydronephrosis/ Cr, 0.7mg/dL
WT	12/1	Gastric reservoir with ureteral Mitrofanoff	425(100)		15	Mild hydronephrosis/ Cr, 0.5mg/dL
CS	7/6	Composite gastric-ileal reservoir with ileal Mitrofanoff	350(130)		25	Mild stable hydronephrosis/ Cr, 0.6 mg/dL

Abbreviation: Cr, creatnine.
*Expected capacity = (age in years + 2) × 30.
+Values for normal bladder compliance have not been established. It is generally accepted that a normal bladder will have a compliance of greater than 20 mL/cm H_2O.

a composite ileum and gastric reservoir. The remaining patient with a composite reservoir has complete incontinence of hemi-Kock augmentation. In the three patients with continent gastric reservoirs the average capacity is 350 mL (range, 300 to 400 mL), with adequate compliance (20 mL/cm H_2O) in one patient, marginal compliance in one patient (15 mL/cm H_2O) and low compliance (9 mL/cm H_2O) in the third patient. The two patients with suboptimal compliance showed mild to moderate hydronephrosis. The third patient with a gastric reservoir with adequate compliance and normal upper tract morphology. The two patients with augmentation gastrocystoplasties have capacities of 250 mL (compliance 10 mL/cm H_2O) and 280 mL (compliance 20 mL/cm H_2O). The patient with lower compliance also demonstrates moderate upper tract hydronephrosis. The patient with a composite ileal-gastric reservoir has both excellent capacity (350 mL) and compliance (20 mL/cm H_2O; Table 1).

Continence Score

A scoring system for evaluation of both gastrointestinal and genitourinary continence was applied to quantify the impact of continent reconstruction on the quality of life (Table 2). After urinary reconstruction with gastric or bowel segments (seven patients) four have achieved a continence score of 3 (intermittent catheterization plus gastrointestinal stoma) and two who had an abdominoperineal pull-through procedure have a score of 4 (intermittent catheterization and fecal control with enema program). Continent reconstruction failed in one patient who has a score of 2 with stomas. The two variant patients

Table 2. Continence score (bowel and bladder)[6]

Voluntary control	3
Control, catheter or enemas	2
Incontinent, appliance	1
Incontinent	0

NOTE: Best score (genitourinary plus gastrointestinal) = 6.

possessed anatomy more favorable than encountered with classic cloacal exstrophy and display continence scores of 5 (volitional voiding plus fecal control with enema program) and 4 (intermittent catheterization plus fecal control with enema program), respectively. The two remaining patients of this series have not undergone reconstruction to achieve urinary continence and are not considered to be appropriate candidates. Their continence scores are 2 (closed exstrophy with epispadias) and 2 (ileal conduit and colostomy, respectively (Table 3).

DISCUSSION

The principles of treatment for the patient with cloacal exstrophy have evolved greatly over the last 15 years. Advances in surgical rehabilitation in concert with advances in nutritional support have yielded both an improved survival rate and an improved quality of life. The urologic aspects of treatment follow the general theme of staged management with an ultimate goal of achieving urinary continence and functional correction of the external genitalia. During the neonatal period, the omphalocele is closed, the exstrophic bowel and bladder segments are separated with closure of the lateral intestinal fissure to reestablish bowel continuity, and a hindgut colostomy is formed. Inclusion of the hindgut as the terminal segment of the intestinal tract offers protection against fluid and electrolyte losses and is clearly superior to the alternative of an end ileostomy.[3,6,10]

Every effort should be made to preserve the appendix or appendices during these maneuvers because they may prove useful in urinary tract reconstruction.[11] The bladder halves are brought together in the midline to form a plate. Primary closure of the bladder is preferred but is pursued only if the child is tolerating the procedure well, the bladder is of sufficient size and elasticity, and the abdominal wall fascia and pubii may be brought together without undue tension. Gender assignment is almost always female and gonadectomy may best be performed at this time to avoid further testosterone imprinting in genetic boys. Delayed bladder closure requires bilateral iliac osteotomy, and in our series was undertaken after 6 to 24 months when nutritional status was optimized. However, we believe that prolonged exposure of the bladder plate deleterious, and there is motivation to secondary closure at the earliest possible time. In no case has cystectomy been performed.

At age 2 to 8 years an effort is made to achieve bowel and bladder control. Our series included one variant patient with an intact urethra who was rendered continent with bladder closure alone and a second variant patient with urethral atresia and covered exstrophy treated with an appendiceal continent catheterizable stoma. Diamond and Jeffs[7] have even reported successful functional closure of the bladder in the cloacal exstrophy patient. Usually, however, achieving continence requires more surgical measures to enhance bladder capacity, bladder compliance, and to provide a catheterizable continence mechanism. In addition, myelodysplasia is frequently present, adding a neurogenic com-

Table 3. Genitourinary and gastrointestinal continence

Patient	Age at time of urinary diversion(yr/mo)	Procedure	Continence status (genitourinary)	Continence status (gastrointestinal)	Total
		Post reconstruction			
CS	5/10	Gastric reservoir with ileal Mitrofanoff, ileal augmentation of reservoir after developed perforation	2 (control with intermittent catether)	2 (pull-through)	4
TW	7/8	Closure of bladder neck, augmentation gastro-cystoplasty, trans-ureteroureterostomy, ureteral Mitrofanoff	2 (control with intermittent catether)	2 (pull-through)	4
BK	6/11	Closure of bladder neck, gastro-cystoplasty, bilateral ureteral reimplant, ureteral Mitrofanoff	2 (control with intermittent catether)	1 (colostomy)	3
WT	7/6	Orthotopic gastric urinary reservoir, uteral Mitrofanoff with perineal stoma	2 (control with intermittent catether)	1 (colostomy)	3
MP	4/8	Orthotopic gastric urinary reservoir, uteral Mitrofanoff with perineal stoma	2 (control with intermittent catether)	1 (colostomy)	3
KR					
		Cloacal exstrophy variants			
JM	3/1	Bilateral ureteral reimplantation, appendiceal Mitrofanoff	2 (control with intermittent catether)	2 (pull-through)	4
TT		Closure of bladder (intact urethra)	3 (volitional voiding via intact urethra)	2 (pull-through)	5
		Not reconstructed			
BD		Secondary closoure of extrophy	0 (incontinent-closed exstrophy)	1 (colostomy)	1
TS		Ileal conduit	1 (ileal conduit)	1 (colostomy)	2

ponent to bladder dysfunction. In some cases a small bladder may be best used for vaginal reconstruction, in which case a urinary reservoir must be provided in toto.

The timing of continence surgery depends on several factors. Bladder augmentation or reservoir construction creates the need for intermittent catheterization, a responsibility for which there must be sufficient parental and social support. It is also advisable to wait until the child possesses the intellectual capacity and manual dexterity to manage regular evacuation of the reservoir without assistance.[12] Most often the child is school aged before these requirements are met and complemented by parent and patient motivation. One patient with an ileal conduit and another with bladder closure alone have not undergone reconstruction because of the absence of these prerequisites. Our youngest patient to receive a catheterizable stoma was 3 years of age but had a constant parental presence and has matured now to the point of assuming responsibility for catheterization. Timing should also allow for bowel adaptation, which is usually denoted by the passage of formed stools[3] and should be performed in the setting of good nutritional status. Limited use of the bowel may be tolerated under these circumstances.

Gearhart and Jeffs[13] successfully used intestinal segments from both the ileum and the hindgut for bladder augmentation without incurring electrolyte problems. However, there is no method for preoperatively determining the amount of bowel that may be safely used without incurring fluid and electrolyte losses.[3] For this reason stomach has been an appealing alternative tissue source in the cloacal exstrophy patient. Additional advantages of stomach include less mucous production, possible decreased incidence of infection, net chloride excretion, which is particularly beneficial in patients with renal insufficiency, and the relative ease of developing submucosal tunnels for antireflux and continence.[14] In our series gastric flaps were a component of every augmentation and urinary reservoir construction.

Techniques for urethral and bladder neck reconstruction such as the Young-Dees-Leadbetter bladder neck reconstruction and use of paraexstrophy skin flaps that were developed in the classic exstrophy patient have been applied to the cloacal exstrophy population. The two patients in our series in whom functional closure of the bladder with bladder neck reconstruction was attempted with a Young-Dees-Leadbetter bladder neck reconstruction did not achieve continence partially because of inadequate urethral resistance. Most reports of bladder neck reconstruction in exstrophy-epispadias patients observe the difficulty in reliably generating the appropriate measure of urethral resistance to ensure continence.[15,17] This often leads to adjunctive procedures such as revision of the vesical neck reconstruction or placement of an artificial urinary sphincter if scar tissue is excessive.[19] Surgical revisions, however, are technically demanding and success is not assured. Furthermore the artificial sphincter carries its collection of risks including cuff erosion, prosthesis infection, and the need for surgical revision or replacement. The Mitrofanoff principle has provided an alternative means of constructing continent access to an augmented bladder or reservoir.[19] Mitchell et al.[19] have demonstrated the spectrum of tissues including ureter, tapered ileum, and gastric tube that may be tunneled to provide a catheterizable conduit in the cloacal exstrophy patient. Our two patients with failed bladder neck reconstructions also demonstrated insufficient bladder capacity and compliance. They were treated with bladder neck closure, augmentation gastrocystoplasty, and ureteral Mitrofanoff catheterizable stoma formation. In the past, sacrificing the bladder neck may have meant loss of endoscopic access. However, current flexible endoscopic equipment allows for inspection via the Mitrofanoff tract should the need arise.

Our urodynamic data suggest that when stomach flaps are used to augment very small bladders or used to construct purely gastric reservoirs the resulting compliance may be suboptimal. Although no decline in renal function has been observed, moderate hydronephrosis has been observed in two patients with gastric reservoirs and one patient with an augmentation gastrocystoplasty. For this reason consideration should be given to formation of a composite reservoir when there is concern about the size of the reservoir that may be constructed purely from stomach. A short length of intestine may be added with the benefit of increasing capacity, compliance, and possibly balancing salt loss.[20] There does not appear to be any detriment to fluid and electrolyte balance when use of small bowel is limited. In fact, there may be complementation of secretory electrolyte loss from the gastric segment and electrolyte reabsorption from the intestinal segment.[21] After developing reservoir perforations twice, one patient underwent ileal augmentation of a gastric reservoir and now demonstrates excellent compliance and has normal electrolytes.

Stoma positioning with urinary tract reconstruction has progressed from invariable use of an abdominal site to an orthotopic urethral location in the selected patient.[9] We also believe that every effort should be made to mimic normal anatomy in reconstructive efforts. In two patients a distal ureteral segment was advanced to the perineum as an orthotopic urethra to provide a catheterizable stoma. This maneuver requires that the patient possess not only good dexterity but sufficient flexibility to easily access the perineum. Both of our patients have a myelomeningocele but are ambulatory. The perineal urinary stoma seems to contribute to a more positive self image.

The goals of surgical reconstruction for the cloacal exstrophy patient are founded on an attempt to approximate normal form and function of the gastrointestinal tract and genitourinary tract. This is of course an idealistic approach that must be tempered by the recognition of anatomic, function, neurological, intellectual, and social constraints as discussed. Innovative techniques in reconstruction of the lower urinary tract have made continent urinary diversion a possibility for the cloacal exstrophy patient who is able to perform intermittent catheterization. Use of gastric flaps is central to these efforts. If the stomach segment that may be harvested for reconstruction is limited in size, however, the use of ileum to form a composite reservoir has been found to be safe and effective.

REFERENCES

1. Rickham PO: Vesico-intestinal fissure. Arch Dis Child 35:97–102, 1960
2. Tank ES, Lindenauer SM: Principles of management of exstrophy of the cloaca. Am J. Surg 119:95, 1970.
3. Howell C, Caldamore A, Snyder H, et al: Optimal management of cloacal exstrophy. J Ped Surg 18:365–269, 1983.
4. Stolar CJH, Randolph JG, Flanigan LP: Cloacal exstrophy: Management through a staged surgical approach. J Ped Surg 25:505–507, 1990.
5. Hurwitz RS, Manzoni AM, Ransley PG, et al: Cloacal exstrophy: A report of 34 cases. J Urol:1060–1064, 1987.
6. Ricketts RR, Woodard JR, Zwiren GT, et al: Modern treatment of cloacal exstrophy. J Ped Surg 26:444–450, 1991.
7. Diamond DA, Jeffs RD: Cloacal exstrophy: A 22-year experience. J Urol 133:779–782, 1985.
8. Lund DP, Hendren WH: Cloacal exstrophy: Experience with 20 cases. J Ped Surg 28:1360–1369, 1993.
9. Mitchell ME, Brito CG, Rink RC: Cloacal exstrophy reconstruction for urinary continence. J Urol 144:554–558, 1990.
10. Hussman DA, McLorie GA, Churchill BM, et al: Management of the hindgut in cloacal exstrophy: Ileostomy versus colostomy. J Ped Surg 23:1107–1113, 1988.
11. Longaker MT, Harrison MR, Langer JC, et al: Appendicovesicostromy: A new technique for bladder diversion during reconstruction of cloacal exstrophy. J Ped Surg 24:639–641, 1989.
12. Gearhart JP, Jeffs RD: Reconstruction of the lower urinary tract. Dial Ped Urol 13::4, 1990.
13. Gearhart JP, Jeffs RD: ,Techniques to create urinary continence in the cloacal exstrophy patient. J Urol 146:616–619, 1991.
14. Adams MC, Mitchell ME, Rink RC: Gastrocystoplasty: An alterative solution to the problem of urological reconstruction in the severely compromised patient. J Urol 140:1152–1156, 1988.
15. Jones JA, Mitchell ME, Rink RC: Improved results using a modification of the Young-Dees-Leadbetter bladder neck repair. Br J Urol 71:555, 1993.
16. Stein R, Fisch M, Stockle M, et al: Urinary diversion in bladder exstrophy and incontinent epispadias: 25 years of experience. J Urol 154:1177–1181, 1996.
17. Allen TD. Editorial: An honest look at bladder exstrophy. J Urol 154:1182, 1995.
18. Perlmutter AD, Weinstein MD, Reitelman C: Vesical neck reconstruction in patients with epispadias-exstrophy complex. J Urol 146:613–615, 1991.
19. Mitrofanoff P: Cystomic continente trans-appendiculaire dans le traitement des vessies neurologues. Chir Ped 21: 297, 1980.
20. McLaughlin KP, Rink RC, Adams MC, Keating MA: Stomach in combination with other intestinal segments in pediatric lower urinary trace reconstruction. J Urol 154:1162–1168, 1995.
21. Lockhart JL, Davies R, Cox C, et al: Acid-base changes following urinary tract reconstruction for continent diversion and orthotopic bladder replacement. J Urol 1542:338, 1994.

32

INDICATIONS AND TYPES OF BLADDER AUGMENTATION IN CLOACAL EXSTROPHY

Michael E. Mitchell

Children's Hospital and Medical Center
Seattle, Washington

ABSTRACT

Cloacal exstrophy is the most severe form of the bladder exstrophy complex. It represents the greatest challenge because not only is the bladder, bladder neck, urethra, and genitalia involved with this deformity and malformation, but the intestine may be severely shortened and innervation to the bladder and bladder neck area may be abnormal because of spinal malformation (present in greater than 75% of these patients). Therefore, reconstruction for urinary continence is a major challenge.

Stomach or a combination of stomach with small bowel may be ideal for bladder reconstruction in these patients. Large bowel or hindgut should *not* be used for the bladder. The advantage of stomach is that it has a structure similar to normal bladder, thus facilitating ureteral reimplantation or tunnels for continence.

INTRODUCTION

For all the challenges that epispadias and exstrophy present the most severe is that of the challenge of the cloacal exstrophy patient. Cloacal exstrophy involves not only the genitalia and urethral plate and bladder, but also usually includes an omphalocele, severe shortening of the hindgut and small bowel and often has associated malformations of the spine and lower extremities. The potential for shortened bowel and neurogenic dysfunction of the bladder and bladder neck make reconstruction in these children a challenge. Although voiding with continence has been achieved in some patients, this is not the rule as usually these patients will be dependent on intermittent catheterization. *However, good urine control with intermittent catheterization should be norm in these patients.*

The spectrum of deformity in the cloacal exstrophy patient can be quite significant and the severity of the omphalocele usually dictates the nature of closure. Small omphalo-

The Exstrophy–Epispadias Complex, edited by Gearhart and Mathews.
Kluwer Academic / Plenum Publishers, New York, 1999.

cele patients can have primary closure of the abdomen and bladder (and more recently of the reconstruction of the genitalia in both boys and girls). Severe omphalocele, however, usually requires a *staged* primary closure because the abdominal wall is usually not adequate to permit complete repair without increased abdominal pressure which risks the abdominal organs and respiratory function. It is sometimes wise to close in staged manner to take advantage of the fact that the posterior bladder halves can be joined with the abdominal wall closure initially, but the bladder itself can serve as a "silo" to permit decompression. Several weeks later the bladder can then be closed appropriately.

Reconstruction of the bladder is usually reserved for a later time, again depending on the nature of the severity of the cloacal exstrophy.

METHODS

The first and primary principle of primary closure in cloacal exstrophy is to preserve all hindgut and tubularize as much as possible of the hindgut that is found open between the bladder halves. Never give into the temptation of permitting this tissue to be used for bladder augmentation primarily. *It is much more important that the hindgut be in the gastrointestinal tract to preserve fluid and electrolyte balance.* The bladder halves can be joined together in a single-stage procedure or, as mentioned above, as a two-staged process depending on the size of the omphalocele. In the single-stage primary closure and in the second-stage of a staged closure effort should be made to move the bladder neck as far posteriorly as possible and this would include in boys reconstruction of the phallus at the time of bladder repair (because of diminutive phallus sexual conversion may be considered, but in general we have avoided this recently and kept male assignments even with rather small phallus development).

It should be noted that the spectrum of the cloacal patients includes *atypical* cloacal patients which, I suspect, may reflect a different pathophysiology and may actually be a twinning phenomenon. The important thing with the atypical cloacal exstrophy patients is that usually there is plenty of hindgut and large bowel. Sometimes a normal component of large bowel which has great implication for the ultimate function and health of these patients.

In the standard cloacal exstrophy patient, excepting those patients who have cycling with continence early on, it is usually necessary to perform a continence procedure at the time of bladder reconstruction. Initial closure of the bladder may be all that is necessary. It has been my experience that the majority of cloacal exstrophy patients need to have bladder augmentation and continence mechanism created. Patients with spine deformity should be presumed to have great potential for neurogenic dysfunction of the bladder neck and urethra and, therefore, tunnel techniques (Mitrofanoff principle) are usually most effective in securing continence.

The type of bowel to use is an issue for discussion because most of these patients tend to have short intestine the colostomy/ileostomy vary considerably in output character. Usually, if the output of the colostomy or ileostomy is liquid, then use of larger hindgut or small bowel should be minimized because of salt and electrolyte loss and poor nutrition.

Stomach becomes an ideal tissue for augmentation of the bladder or creation of a continent reservoir.[1,2] Sometimes this may be combined with a small segment of bowel, but minimizes the need for extensive use of small bowel.

The implications of intestine or stomach in the lower urinary tract in the cloacal exstrophy patient is quite significant. Removing small intestine, or certainly large bowel, from the gastrointestinal tract in these patients reduces resorption potential, but in the same sense these patients tend to lose salt and electrolytes through the gastrointestinal tract. If stomach is used in the reconstruction of the bladder, salts and acids may be lost from the gastric segment. The net conclusion may be that it is probably better to use a smaller segment of stomach and small bowel and make a composite reservoir which would help to maintain total salt balance. The rule of thumb, though, in these complex reconstructions is that multiple alternatives should be available and the reconstruction guided by what tissues seem to be most appropriate after dissection is complete.

A significant challenge in these patients is also presented not only in the selection of the material for bladder reconstruction, but also in the technique for constructing a continent channel for dependable catheterization. We have utilized the ureter in an orthotopic position in many of these patients.[3] This is facilitated by the fact that the ureter in the cloacal exstrophy patient tends to be elongated and a distal segment can be used to construct a catheterizable channel. Recently we have used a technique in which the ureter is kept intact and tunneled through the bladder and brought out in the perineum such that the blood supply to the ureter is preserved. The ureter is basically tunneled twice, into the bladder and then out the base of the bladder such that there is a tunnel for continence and a tunnel to prevent reflux and the side of the ureter is opened between. Recently proposals have been made for a more dependable tapered small intestinal segment to be used in tubes to secure continence. Because the cecum is almost never available, nor is appendix in these patients, ileocecal valve and nipple reconstructions seem to be not often appropriate.

CONCLUSION

The broad spectrum of the cloacal exstrophy patient precludes a single effective strategy for reconstructions. Because many of these patients have altered innervation and minimal tissue to work with they remain a challenge for reconstruction. Nevertheless, the majority of these patients can be reconstructed in a manner which will permit urinary control with intermittent catheterization. Stomach or stomach in combination with small bowel can be very effectively employed in reconstruction in these patients.

REFERENCES

1. Adams, M.C., Mitchell, M.E. and Rink, R.C.: Gastrocystoplasty: An alternative solution to the problem of urological reconstruction in the severely compromised patient. J. Urol., 140:1152, 1988.
2. Mitchell, M.E., Brito, C.G. and Rink, R.C.: Cloacal exstrophy reconstruction for Urinary Continence. J. Urol., 144:554, 1990.
3. Sumfest, J.M., Burns, M.W. and Mitchell, M.E.: The Mitrofanoff principle in urinary reconstruction. J. Urol., 150:1875, 1993.

33

CLOACAL EXSTROPHY

Experience with 41 Cases from 1976–1997

W. Hardy Hendren,[1] Robert E. Gross,[2] and Dennis P. Lund[1,2]

[1]Children's Hospital
Boston, Massachusetts
[2]Harvard Medical School
Boston, Massachusetts

INTRODUCTION

Cloacal exstrophy is one of the most formidable challenges in the field of pediatric surgery. It is said to occur with an incidence of 1 in 200,000,[1] to 1 in 400,000 live births.[2] Thus, in the United States with a birth rate of about 3 million per year, there should be at most only 1 to 2 dozen such babies who are live born each year. It would seem prudent, therefore, that the treatment of these complex problems be directed to a limited number of centers where there are surgeons with special interest in caring for these relatively rare problems.

Cloacal exstrophy is thought to be due to premature rupture of the cloacal membrane prior to descent of the urorectal septum.[3] Principle features of cloacal exstrophy include omphalocele, exstrophy of two hemibladders, an open cecum between the two hemibladders, microcolon and imperforate anus, and ambiguous genitalia (Figure 1). Orthopedic[4] and upper tract urinary anomalies are frequent. Rickham reported the first survivor relatively recently, in 1960.[5] Our own experience with cloacal exstrophy, like that in other centers in the 1950's and 1960's was dismal. Death occurred from fluid and electrolyte losses, malnutrition, and sepsis. Today nearly all of these babies can survive, based on improved methods in their management, modern pediatric anesthesiology, and surgical techniques which have developed in the past few decades.[7–11]

Although the first sight of a newborn with cloacal exstrophy may prompt thoughts of futility and disappear, in many of these babies proper management can produce a happy and productive little citizen for the future. This chapter will present our current thoughts regarding management of cloacal exstrophy, with description of illustrative cases, and in particular with reference to follow up after reconstructive surgery.

The Exstrophy–Epispadias Complex, edited by Gearhart and Mathews.
Kluwer Academic / Plenum Publishers, New York, 1999.

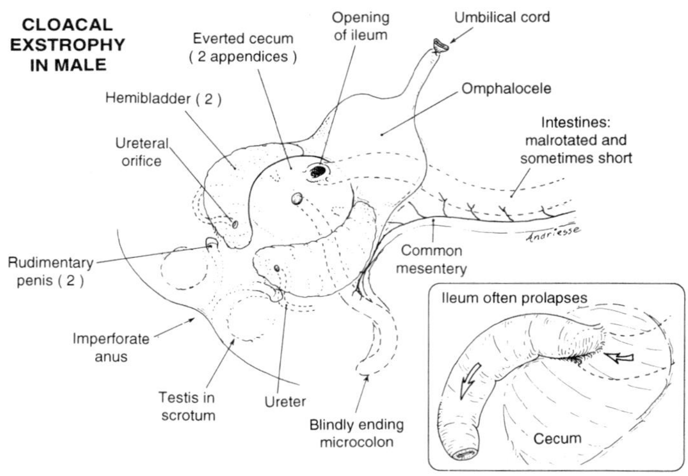

Figure 1. Principle anatomic features of cloacal exstrophy in a male baby. Note everted cecum between two hemibladers connected to the cecum, a blindly ending microcolon, which shares a common mesentery with the terminal ileum. The ileum often prolapses. Integrity of perineal muscle is excellent in some patients and completely deficient in others.

Case Material

From 1976 through 1997 we treated 41 patients with cloacal exstrophy. Eleven were new primary cases, and 30 were secondary cases who had undergone some type of reconstructive operation before referral. Twenty-four were genetic males and 17 genetic females. Only 4 of the genetic males had been reared in the male role. One of these was an adult, very distraught about his completely inadequate phallus. Another was an infant born to a family with 3 female siblings, whose parents were adamantly opposed to rearing him in the female gender role. Two were cloacal exstrophy variants who had a reasonable phallus and who could, therefore, regard themselves as males

Years Postoperatively

One patient is 21 years after the beginning of reconstructive surgery as an infant. Ten patients are ten years or more postoperatively. Fourteen are between 5 and 9 years later. Twelve are less than five years following reconstructive surgery. Five are currently preoperative patients, having undergone initial neonatal operations but not subsequent definitive surgery for continence.

Augmentation Cystoplasty

Many of these patients require enlarging bladder.[12] Twenty-four of our patients had bladder augmentation. Four had two augmentations (stomach and small bowel); 11 had gastrocystoplasties; 9 had augmentations with small bowel. There are two patients, whose bladder was of adequate size, requiring no augmentation. Four patients have a urinary diversion. In 11 patients the urologic aspect of their reconstructive surgery has not been completed.

The Bowel

Eighteen patients had abdominal perineal pullthrough of the short colon, but in two this was subsequently reversed. One of those two toddlers ingested a wire "baggy tie" which perforated the rectum, causing pelvic phlegmon, and need to return to a colostomy. A second child underwent colostomy elsewhere, which we believe was premature because not enough effort had been spent trying to control loose stools medically with diet and medication. Of those who had pull through, six have spontaneous bowel movements and are clean between movements. However, the majority of these patients require an enema washout program to evacuate the colon.

The Bladder

Six patients have a reversed nipple as their continence mechanism.[13] In one case the nipple required revision 12 years later because the patient became unable to pass a catheter easily. Trimming and straightening the nipple cured the problem. A 7th patient had a nipple, which sloughed, and so an alternate form of continence was constructed (Mitrofanoff abdominal stoma from tapered small bowel). Four patients have a continent diversion reservoir which is emptied by self-catheterization through a small abdominal stoma. Four patients have a urinary diversion, and wear a collecting bag on a stoma. Sixteen patients had urethral reconstruction. The majority of them self catheterize (13); 3 are able to void by straining. Eleven patients have not yet had urologic reconstruction for continence.

ILLUSTRATIVE CASES

Case 1

LK was referred at birth with typical features of a genetic male with cloacal exstrophy (Figure 2). At the hospital where the baby was born a suggestion had been made to offer no treatment, but the family did not wish to pursue that course. Before performing the neonatal operation the gender for rearing was discussed thoroughly with the family. They elected to follow the advice to raise the baby as a female, because there was inadequate phallic tissue available from which a satisfactory penis might be constructed later.There was a solitary right kidney in normal position and bilateral clubfeet. On the first day of life the infant underwent an 8 hour procedure. The cecum was separated from the two hemibladders, closing the cecum, performing a decompressing loop ileostomy superior to it. The blindly ending microcolon distal to the cecum, and measuring 2 ½ inches in length, was pulled through to the perineum and placed in the sphincter muscle, which reacted briskly to electrical stimulation. The two hemibladders were joined together and closed. Bilateral orchiectomy was performed. The small omphalocele was closed. The pubic symphysis was approximated and the baby was placed in modified Bryant's traction.

The tethered spinal cord was released three months later by Dr. Michael Scott.[14] The ileostomy was closed at age 7 months, after demonstration of a satisfactory lumen of the cecum and the colon distal to it. Rectal dilatation has been performed meanwhile to prevent stenosis of the pulled through microcolon while it was defunctioned. Saline irrigations had been done also via the anus and out the distal limb of the loop ileostomy to encourage lengthening of the defunctioned colon. Bilateral inguinal hernias were repaired

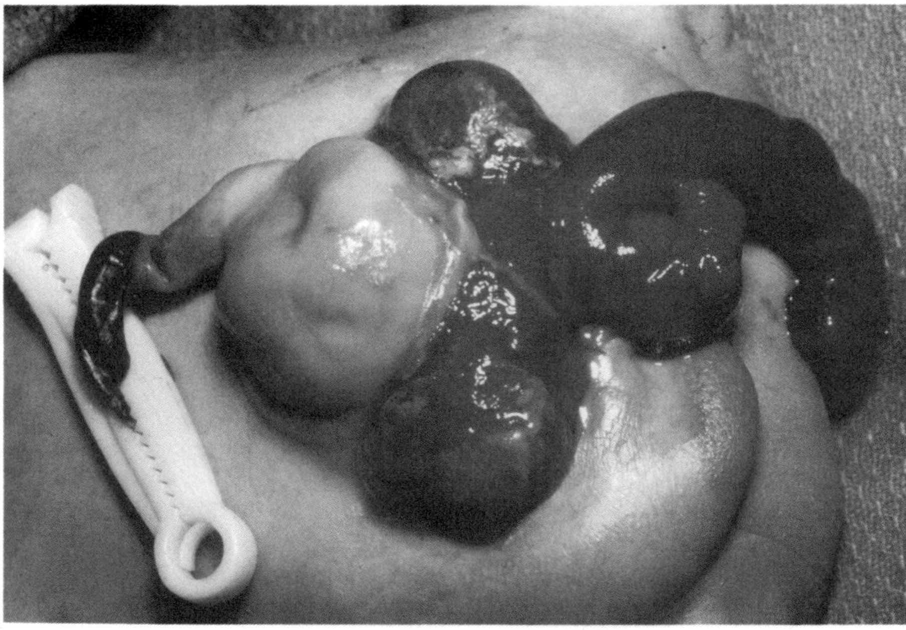

Figure 2. Photograph of patient 1 at birth. Illustrating all of the features shown in Figure 1. There is a relatively small omphalocele, two hemibladders, eversion of the cecum and prolapse of the terminal ileum like a "elephant trunk". Operation was performed as shown in Figure 3A.

at age 13 months. The bladder was incontinent, as expected, since no procedure had been done to achieve continence at birth. Cystogram showed reflux and a small bladder.

At age 2 ½ years a reconstructive operation was performed consisting of reimplantation of the right ureter, creation of a reversed ileal nipple valve for urinary incontinence, augmentation of the bladder with stomach,[15] and creation of a small bowel vagina. Unfortunately the ileal nipple sloughed due to ischemia, and so the operation was a failure as regards urinary continence. Her anatomy after healing of the sloughed outlet nipple is shown in Figure 4. Nineteen months later, at age 4 years, another reconstuctive operation was performed as shown in Figure 5. Decision was made to opt for a continent urinary reservoir to be catheterized through a small right lower quadrant stoma. The bladder outlet was closed. A catheterizable Mitrofanoff stoma was made by plicating the prior small bowel vagina into a catheterizable conduit and implanting that into the back wall of the gastrocystoplasty. A new implantation of the solitary ureter was performed while making the Mitrofanoff conduit. The bladder size did not seem adequate despite prior augmentation, possibly because it had been freely incontinent. Therefore, a second augmentation using small bowel was added to it. Another small bowel vagina was created. This complicated and difficult procedure lasted 18 hours. It was successful. The patient has been completely dry. Her mother catheterizes with a small curved metal cannula. A daily soap suds enema is given to empty the colon, which is continent between washouts. The microcolon, which had been only 2 ½ inches long initially, is now 30–40 cm. In length and functions well as a colon, with appropriate water absorption and fecal storage between washouts. This child functions well at age 9 years. She has required surgery for her legs, but she is ambulatory with braces and crutches. She interacts well with two older male siblings. She is a focus of great love and attention within the family. She has won multiple gold medals

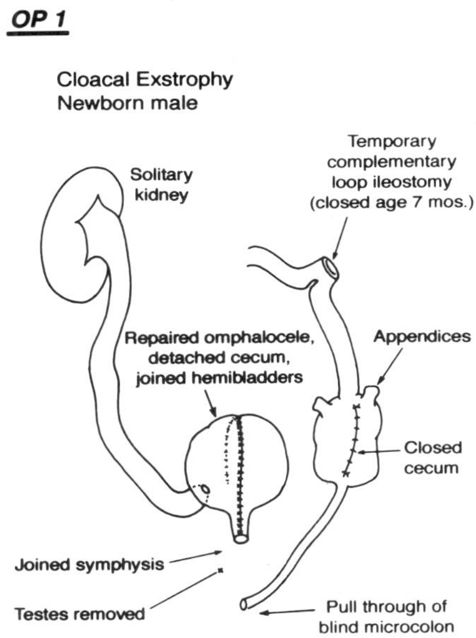

Figure 3A. (Case 1) Anatomy after initial operation showing pulled thru colon and solitary kidney. Note protective complementary ileostomy which was closed later.

in New York State Special Olympics in competition with 1200 other children with special needs. She is bright, energetic, and well motivated. It is predictable that she will succeed with something worth while in life.

Comment

This child illustrates a number of important points. First she underwent an 8 hour operation as a neonate and two subsequent long reconstructive operations. The importance of expert anesthetic management can not be overemphasized. The evolution of today's superb anesthetic management has been essential for performing such global reconstruc-

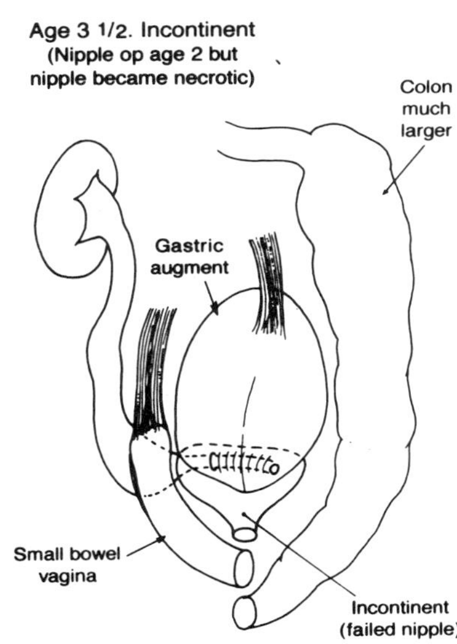

Figure 3B. Anatomy after initial reconstruction at age 3½ years. A nipple valve was constructed for continence but became ischemic and disappeared. Note growth of colon after closure of complementary ileostomy at age 7 months. Small bowel vagina located between bladder and rectum. Bowel function was excellent at this point, with a daily enema washout. The patient remained incontinent of urine.

OP 3

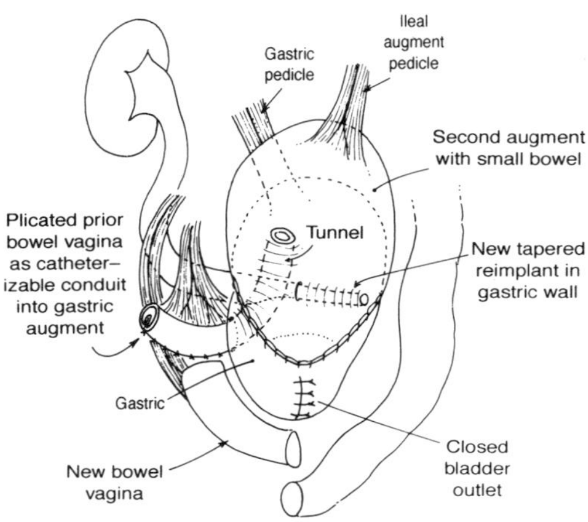

Figure 3C. Anatomy after 3rd major operation, with closure of bladder outlet, construction of a new bowel vagina, tapering of the former bowel vagina to make a catheterizable conduit implanted into the gastrocystoplasty, second augmentation with small bowel, repeat reimplantation of ureter.

tions. Secondly, this case illustrates the principal of not giving up the quest for a satisfactory solution after an initial failure. There is almost always some alternative to simply giving up and opting for a permanent urinary diversion into a bag on the abdomen. That course is easy for the surgeon, but not so easy for the child. Persistence on the part of the surgeon in these complex reconstructive cases is essential to obtain continence, which is true in many pediatric urologic conditions.

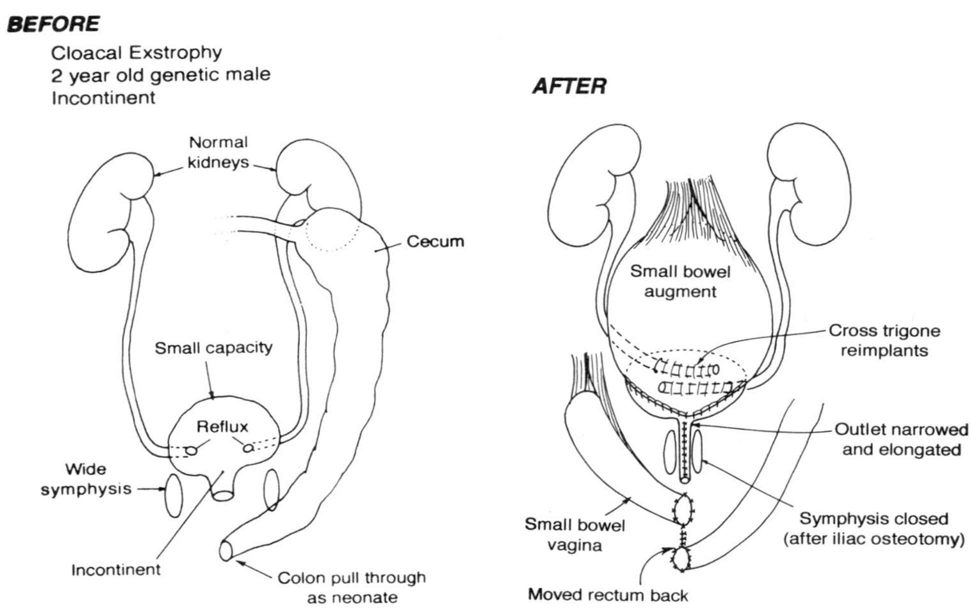

Figure 4. (Case 2) (Left) Preoperative anatomy showing patulous bladder outlet and normal kidneys with reflux. The colon had been pulled thru as a newborn. (Right) After reconstruction showing narrowed bladder neck, cross trigone ureteral reimplants and small bowel augmentation. The rectum was moved back to make room for a small bowel vagina.

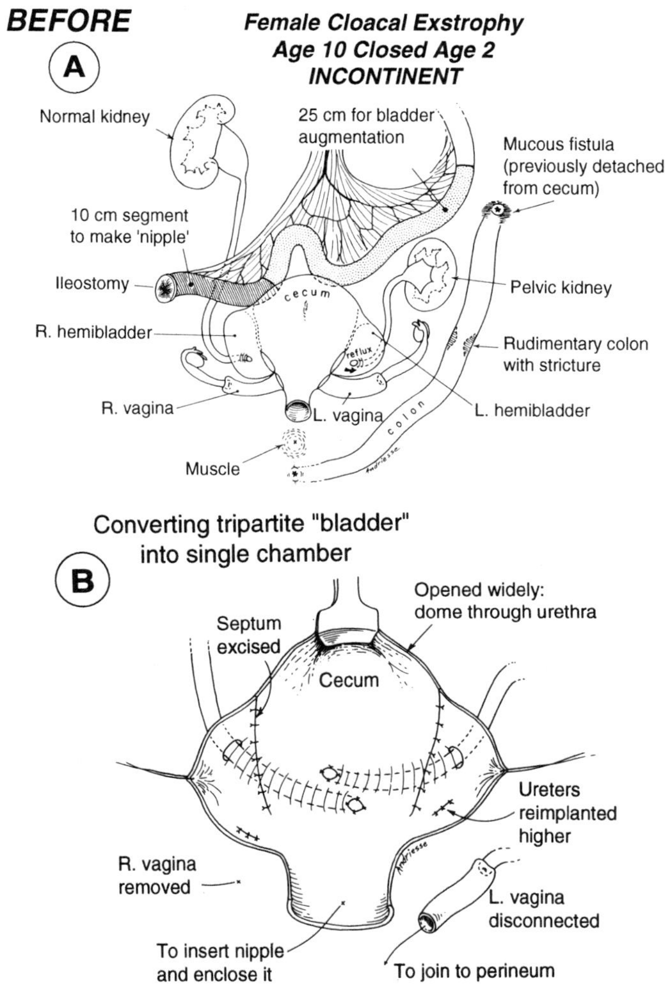

Figure 5. (Case 3) A) Preoperative anatomy showing tripartite bladder which consisted of two small hemibladders and the cecum. There was a colon stricture which precluded use of the hindgut in the gastrointestinal tract. B) opening the bladder to make a single chamber, disconnecting the left vagina and removing the right vagina and uterus. Note: ureteral reimplant.

Case 2

(RL) was referred at age 26 months (Figure 4a). Antenatal diagnosis had disclosed an abdominal wall defect. Surgery as a neonate included separation of the gastrointestinal tract from the bladder, closure of the bladder and omphalocele, and primary pullthrough of the colon to the perineum. Testicles were removed and the patient was raised as a female. She had excellent bowel function with a daily enema washout. The bladder capacity was small. There was vesicoureteral reflux bilaterally to normal kidneys. At age 34 months, urologic reconstruction was performed. This included iliac osteotomies to approximate the soft tissue of the bladder outlet more easily, cross trigone reimplantation of the ureters, augmentation of the bladder with ileum, creation of a small bowel vagina, and moving the rectum back more posteriorly. One month later an emergency operation was required for

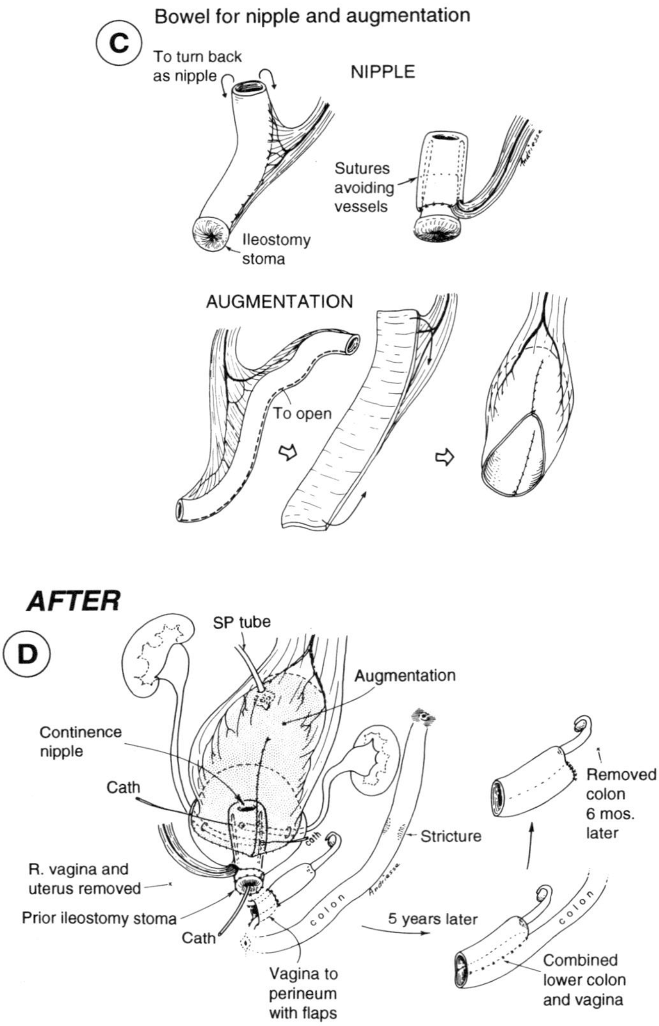

Figure 5. C) using small bowel to create a nipple valve for continence and an augmentation to increase the reservoir capacity of the bladder. D) completion of reconstruction for continence. Subsequent augmentation of pulled through vagina using distal colon, which was deemed not usable because of colitis and stricture.

gastric volvulus. Seven months later a repeat operation was needed to narrow the bladder outlet further because the patient as still wet. The patient is now 13 ½ years old. She is completely dry and self catheterizes every four hours without difficulty. Her urinary tract is stable. She is maintained on suppressive antibiotics to prevent bacilluria in the augmented bladder. Biofeedback training two years ago has enabled her to empty the colon without the daily enema washout, which is now done about once a week to assure that she is not becoming impacted with stool.

Comment

This child illustrates what can be accomplished with a lot of perseverance on the part of a child and a devoted mother. The patient and her mother have served well as a counselor to other children with cloacal exstrophy, serving as an example that some of these children can lead an essentially normal life with no stoma on the abdomen. She looks like a female and behaves like one. To date in this child there has been no hint that raising her as a female despite the opposite genetic gender at birth has presented a problem. Estrogen therapy has not been started yet, in order to maximize skeletal growth while her stature is smaller than her peers

Case 3

(DK) was referred at age 10 years for possible correction of urinary incontinence. She was a genetic female with a pelvic kidney on the left, and a normotopic kidney on the right. (Figure 5A). The hindgut had been separated from the cecum at birth, and exteriorized as a mucous fistula in the lower abdomen, and a perineal pullthrough of its distal end. Having never been functionalized, the colon remained small. There was a colon stricture secondary to possible colitis. The vaginas entered the bladder. There was vesicoureteral reflux. Urine drained constantly from the patulous bladder neck. The perineum and buttocks were severely excoriated as a result. At operation the bladder opened widely, putting a sound through the perineum to the dome of the tripartite bladder which was cecum and laying it open completely before deciding what might be possible. [16] The vaginas were widely separated. Therefore, the decision was made to pullthrough the left one, with its uterus, removing the right one. The two hemibladders and cecum were fashioned into a single viscus by opening the septum between them. There did not seem to be adequate soft tissue to roll in as a urethra and bladder neck. Therefore, the distal 10 cm. of ileum, which had functioned as ileostomy since infancy, was turned back as a nipple to place into the bladder outlet. This was the first case in which we tried that method. The next segment of ileum was fashioned into an augmentation cystoplasty of small bowel. Figure 5 C and D. The previously pulled through colon was used later to augment the caliber of the pulled through vagina. The vagina now accepts a #26 hegar dilator. She has been menstruating normally since age 16 years and may have the potential for becoming pregnant in the future. The patient, now age 23 years is a college graduate, has excelled in the sport of swimming, and there is little doubt that she will have a successful career in her chosen field. She has a nonidentical twin sister who had no malformations. The patient does not suffer by comparison, in terms of academic and social success.

Comment

This patient illustrates that in the neonate with cloacal exstrophy every attempt should be made to salvage all colon tissue to function as colon. The cecum, an important part of that, had been left with the hemibladders to provide extra reservoir capacity for future urologic reconstruction. Meanwhile the microcolon withered and developed colitis without stricture. To use it many years later did not seem feasible in view of the stricture. Therefore, she must continue with an ileostomy. Fortunately she is well adjusted to that, having known no other state since infancy.

Case 4

(AL) was referred at age 12 years for creation of a urinary continence mechanism. She was known to have a short small intestine by description of the referring surgeon and also evident on a barium swallow with small bowel follow through. There was a large ventral

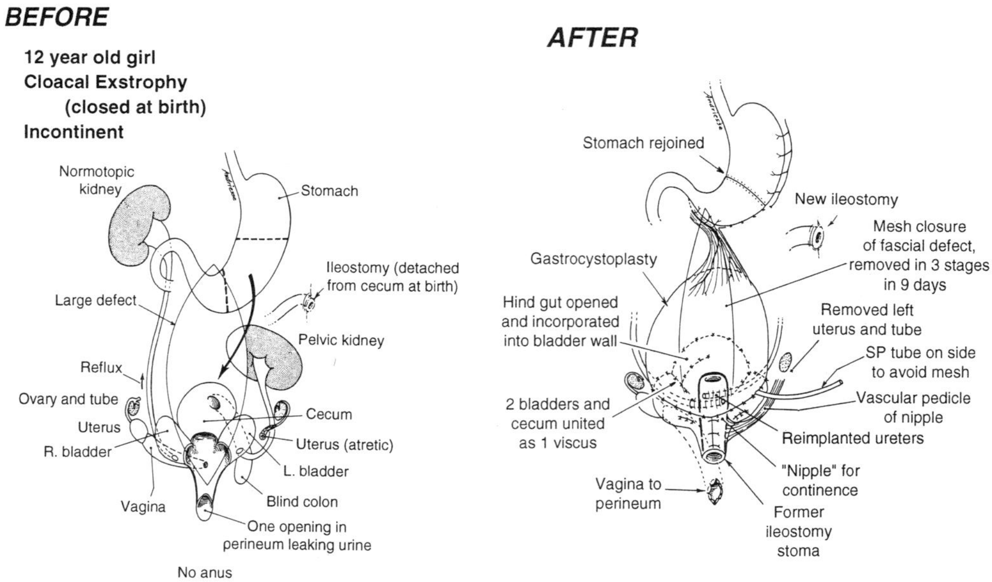

Figure 6. (Case 4) (Left) Anatomy when referred for operation to attain urinary incontinence. Cecum and hindgut were still attached to the bladders. (Right) After augmentation using stomach and creation of nipple for continence. Cecum and hindgut were retained to provide additional reservoir capacity. In retrospect, it might have been better to attempt adding them to the G.I. tract, with the possible gain of eliminating the ileostomy.

hernia which extended from the epigastrium to the pubic region. The right kidney was normotopic and the left in the pelvis.There was uterine atresia on the left. There was free vesicoureteral reflux. Gastrocystoplasty appeared to be the only reasonable way to augment the bladder which already included all of the colon. Deficient length of small bowel ruled out use of intestine as an option. Detaching the cecum and salvaging the hindgut was considered, but not elected. In retrospect that might have been a better course to pursue. The anatomy was reconstructed as shown in Figure 6, on right. A nipple was used for the continence mechanism, using a short segment of terminal ileum for that. The right vagina was disconnected and pulled through to the perineum. The atretic left uterus was removed. Gastrocystoplasty was performed to enlarge the bladder. The large ventral hernia could not be closed primarily. Mesh was placed and removed in three stages, during the next 9 days, as would be done in a newborn with a large omphalocele with insufficient room in the abdomen for primary closure. Delayed primary closure of skin and subcutaneous fat were performed after fascial closure of the abdominal wall. The patient, once in diapers with total urinary incontinence, made a complete reversal in her social status. She is currently age 22 ½ years, in college, and enjoys a greatly improved self image. An ovarian cyst on the left side recently required aspiration. Menstrual periods are present, but irregular. She is dry.

Comment

This patient like case 4, underscores the need to separate the GI tract from the urinary tract in all of these patients, to salvage the colon for improved gastrointestinal function and to consider the possibility of doing a pullthrough, which was not considered feasible in the past. We wonder in retrospect if it might have been better to do a late sepa-

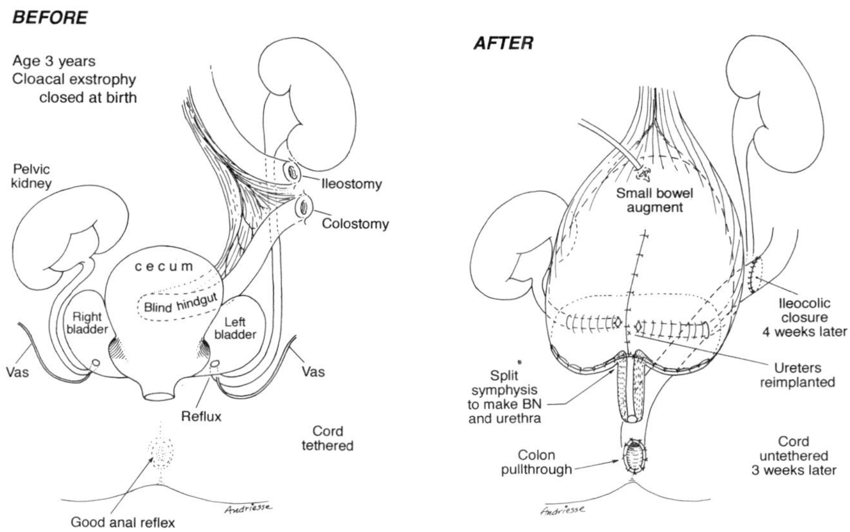

Figure 7. (Case 5) (Left) Anatomy before reconstruction. (Right) Scheme for reconstruction for continence of urine and stool, both of which were accomplished.

ration of the cecum and blind ending colon as a separate procedure, to see if bowel continuity could be established before embarking on the urologic reconstruction. It is of vital importance in these complex cases to consider all options possible, before "burning any bridges", such as using precious colon in the urinary tract where other options include small bowel and stomach.

Case 5

(RP) was referred at age 3 years for possible reconstructive surgery. At birth the omphalocele has been closed and the gut had been exteriorized as an ileostomy and colostomy, retaining the cecum between the hemibladders (Figure 7, left). There was good perineal muscle. An incidental finding at surgery was entry of the vas deferens on each side into a ureter, which we have encountered in several other patients. Recently we encountered a female with cloacal exstrophy, where the uterus on each side entered a ureter Reconstruction included cross trigone implantation of the ureters for reflux, tubularizing the tissues at the outlet to make a bladder neck and urethra for continence, and pullthrough of the hind gut as anus and rectum. Ileocolic closure was performed later. She is continent of urine on self catheterization and continent of stool, with daily saline enemas to wash out the bowel. She has a tendency to form bladder stones. A large one was removed by open cystotomy, to extract it intact. She drinks a lot of water, and catheterizes on a schedule to discourage further stone formation. On two occasions a small stone has been extracted endocopically, using a stone basket passed through the urethra via an endoscope.

Comment

This is an exuberant outgoing little child, now age 6½ years who takes charge in the school setting with her peers. Although she dresses like a little girl, and considers herself to be a normal little girl, her "take charge" behavior in a group may reflect a male behavior imprint by her male chromosomes. It is too early, at age 6½ to be certain of that. Again, this patient illustrates that when these children are explored at birth to close the omphalocele and separate the GI tract from the urinary tract, every effort should be made to maintain the integrity of the ileum, cecum, and rudimentary microcolon.

Case 6

(AT) was referred at age 12 years with total urinary incontinence and a left upper quadrant end colosotmy, which was stenotic. The bladder was scarred and small. A left vagina entered the bladder. Reimplants had been performed previously, but there was reflux bilaterally. There was absence of a right vagina and uterus. There was right hydronephrosis secondary to the retrocaval ureter. A global operation was performed, seen in Figure 8B. Right ureteral obstructon was corrected by excising the area of stenosis and performing a spatulated anastomosis anterior to the vena cava. The small scarred bladder had no volume. Most of it was used to create a urethra and bladder neck. The outlet mucosa was scarred. Therefore, a buccal graft was used to line the urethra and bladder neck. Augmentation cystoplasty was performed using stomach. The ureters were reimplanted into the stomach as illustrated. The left vagina was pulled through to the perineum. The stenotic colostomy was taken down, tapered, and pulled through to the center of the perineum which had excellent sphincter muscle. The need for blood, albumin solution and crystalloid were enormous, to maintain metabolic homeostasis and a good urinary output during this 27 hour anesthesia. Sequestration of fluid in the bowel wall made impossible primary closure of the abdomen at the end of this marathon procedure. Silastic sheet material was used to close the abdomen, similar to that used in closing a large omphalocele in a neonate. The patient returned to the operating room every two days during the next week, for removal of this prosthetic material. Fascia closure was obtained at the third procedure. Delayed primary closure of the subcutaneous fat and skin followed.

The patient is now age 15 years. She is dry, using intermittent self catheterization. She is continent of stool, using a daily enema washout. She has undergone subsequent surgery for her hips and her back. She is fully ambulatory with the help of crutches, and functions well socially. She aspires to be a physician. We think she will likely accomplish that goal.

Comment

It has been a remarkable social transformation for this little girl who came in diapers, totally incontinent of urine and who was resigned to a lifetime with a colostomy bag. She now has no diapers and there is no bag. The contributions of modern pediatric anesthesia and intensive postoperative care are vital in being able to accomplish an operation of this magnitude. The role of the operating room nurse, Dorothy Enos, RN, and surgical house staff, Matthew Rosenberg, M.D. and Peter LaVesque, M.D. were vital to the success of this operation. They were all present from the start to the end of the operation.

DISCUSSION

Patients with cloacal exstrophy, like patients with "ordinary cloacal malformations" vary in details, from one case to another. Nevertheless experience during the past two decades has given us certain guidelines which are useful.

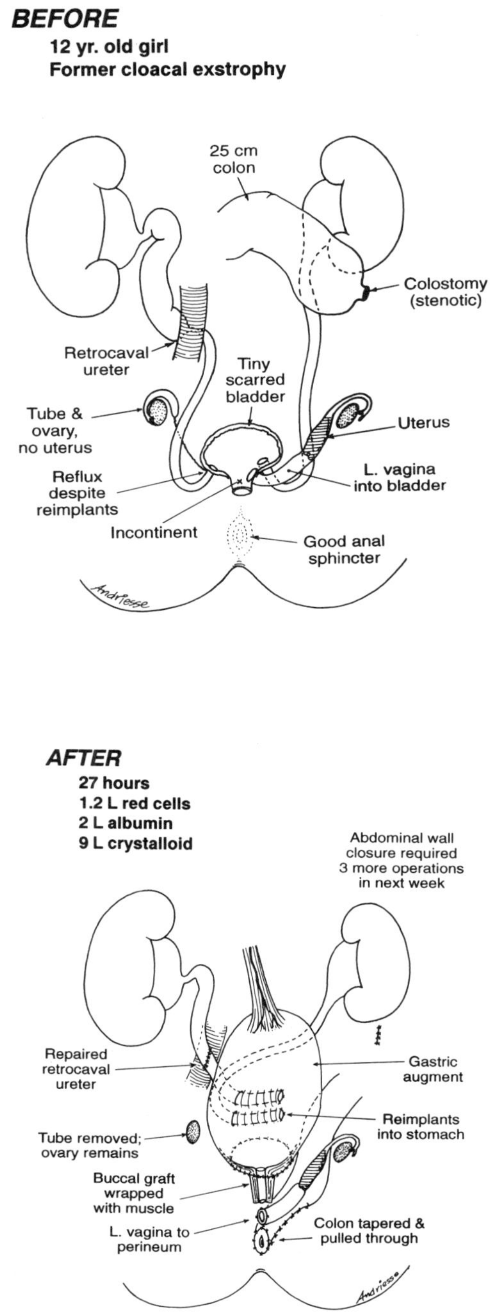

Figure 8. (Case 6) (Top) Complex preopeative anatomy in patient with "permanent" colostomy, and total incontinence of urine. (Bottom) Scheme of global reconstruction which achieved dryness and stool continence. The patient self-catheterizes to empty bladder, and uses enema washouts to empty colon.

The surgeon faces the goals of providing urinary continence, bowel continence and sexual function for the patient in future years. If the infant has a very large omphalocele, neonatal care may be initially to deal with only the omphalocele, deferring all else temporarily. Fortunately prolonged total parenteral nutrition makes possible options which were not present in the 1960's. If the omphalocele is not formidable, neonatal management can include not only omphalocele closure, but also separation of the gastrointestinal and urologic tracts. The two hemibladders should be sewn to one another to form a single bladder.

Iliac osteotomy can help with soft tissue approximation. In the first few days of life bringing the pubis together may be possible without osteotomy, whereas iliac osteotomy is usually needed later. Using modified Bryant's traction for the legs for three weeks can help healing without dehiscence of the wound and bladder closure. We use "mole skin" traction on the legs. In older patients the position can be maintained by passing Kirschner wires through the lower femurs, to suspend the patient in traction. Pubes that are brought together after osteotomy will often drift apart later, so that an interpubic ligament of scar remains rather than bony union. For this reason wire should *never* be used to approximate the pubes, because wire can transect the bladder neck and urethra as the pubis spreads and soft tissues move forward. We use absorbable sutures of heavy polydiaxone.

The rudimentary microcolon should never be removed, because it has the capacity to enlarge, lengthen, and function as colon. It is not uncommon to see a microcolon lengthen to 30 to 40 cm. The decision at birth needs to made whether the microcolon should be exteriorized as a colostomy or pulled through to the perineum. In either case, closing the cecum and bringing the end of the microcolon to the surface of the abdomen, or to the perineum, should be accompanied by a temporary complementary loop ureterostomy, to protect the closure of the cecum. Immediate pullthrough can be considered for a baby with good perineal muscle and no obvious defect like a sacral myelomeningocele. The presence of a good gluteal cleft and muscle which responds well to stimulation would make it reasonable to consider an immediate pullthrough. We see no obvious advantage to a posterior sagittal approach for this if the contractile muscle of the perineum is observed closely and the pullthrough is done exactly in its center point. Today in most pullthroughs for cloacas and imperforate anus, we employ the posterior sagittal approach.

The infant male with cloacal exstrophy usually has a rudimentary hemipenis on each side. The very small size and wide distance between these parts makes subsequent reconstruction of a good phallus unlikely if not impossible in most. There were four males in this group raised as males. Two were cloacal exstrophy variants where a reasonable phallus was present, which is unusual. One adult had been raised as a male elsewhere, because he was genetically male and had male gonads. He remains very unhappy, confused, and frustrated having male sexual desire, but anatomy incompatible with that gender role. The second is a genetically male infant with three sister siblings, whose father is adamant about raising him as a male. Dissuading the family from that course was not possible. The ultimate outcome will not be evident for many years. There is no ideal manner to manage these genetically male babies. The same is true for many types of intersex problems. In general we have felt that the incomplete female can cope better socially than the inadequate male. We do not agree with the recently suggested course of following the genetic gender assignment and allowing the patient choose a gender role in adult life. There is no question that some patients who are reared as females, but who are genetically males, have a definite male imprint despite removal of gonads at an early age and up bringing in the female role. Some definitely become "Tom Boys", and are proficient in sports usually considered to be male activities. It will be important to follow these patients for many years, to learn more of their ultimate sexual identity and behavior. The oldest patient in our series who is 21 years following reconstructive surgery was a genetic male, raised in a very supportive family as a female. She has functioned well with appropriate endocrine support at the time of puberty and creation of a small bowel vagina with which coitus has proven satisfactory. Her true genetic back ground has never been questioned by the patient. The family has chosen, wisely, not to bring up the issue with her. There are some of these patients who may exhibit sexual confusion as adults. There are others who make the gender transition with success.

After the initial surgical management of the newborn with cloacal exstrophy, much remains to be done. The tasks that lie ahead include: 1) Creating a mechanism for urinary incontinence. The bladder nearly always requires augmentation. Choices will be to use small bowel or stomach, because colon is in short supply and should be saved for its important gastrointestinal function. 2) Performing vaginal reconstruction. 3) Possibly performing a rectal pullthrough if it was not done initially. Rectal pullthrough is not appropriate in a patient with a "rocker bottom", where there is no contractile muscle.

The pelvic outlet in these patients is different from that in patients with imperforate anus or cloacas. Therefore, when performing a delayed rectal pullthrough in these patients, We have not used the posterior sagittal approach which is so useful in patients with essentially normal pelvic anatomy. The point of maximum sphincter muscle contraction is identified by electrical stimulation. The dissection is carried straight in through the contractile muscle complex, reaching up with a clamp and pulling through the rectum into the center of the contractile muscle.

It is much more difficult to reconstruct a continent bladder neck and urethra in patients with cloacal exstrophy, than in patients with ordinary bladder exstrophy. The tissues are poorly developed. To make a urethra and bladder neck we split the tissue widely open, creating a strip of mucosa to fashion into a tube and enclosing it with the surrounding muscle to provide outlet resistance. The laterally placed ureters uniformly reflux. They require reimplantation to stop reflux as well as move them upward to allow fashioning a urethra and bladder neck. The back wall of the bladder has a previous suture line where the two hemibladders were sewn together at birth. This can complicate reimplanting the ureters, especially if the bladder is small. An alternative is to reimplant the ureters into an augmentation, especially if it is stomach. It is easier to implant ureters into stomach, where a nonrefluxing tunnel is easily constructed, than into a small, scarred native bladder.

Although the results of reconstruction in patients with cloacal exstrophy are far ahead of where they were 30 years ago, there is still much to be desired. Most of those with a rectal pullthrough will require a washout with saline or soap suds enemas to empty the colon. Although we can make an internal urinary reservoir, most of these patients will need to intermittently catheterize whether it is in orthotopic position or there is a catheterizable stoma. Either of those options is better than incontinence and wearing diapers, or wearing a bag on the abdomen.

It is encouraging that a sizable proportion of these patients with cloacal exstrophy can now be reconstructed so that they wear no bag for either urine or bowel. Although much progress has been made since 1960, there is much more to be done in this type of reconstructive surgery.[17] In America where our birth rate is about 3 million per year, and there are relatively small numbers of these patients, it would seem prudent to concentrate their care in a small number of centers. That will allow surgeons with a real interest in this difficult sort of work to become proficient in performing these reconstructions.

REFERENCES

1. Tank ES, Lindenauer SM: Principles of management of exstrophy of the cloaca. Am J Surg 119:95, 1970
2. Ziegler MM, Duckett JW, Howell CG: Cloacal estrophy. In: Welch KJ,Randolph JG, Ravitch MM et al(eds): Pediatric Surgery. Year Book, 1986, pp 764–77,
3. Thomalla JVRudolph RA, Rink RC et al: Induction of cloacal exstrophy in the chick embryo using the CO2 laser. J Urol 134:991–995, 1985.
4. Greene WB, Dias LS, Lindseth RE et al: Musculoskeletal problems in association with cloacal exstrophy. J Bone Joint Surg (AM)73:551–560, 1991.
5. Rickham PP: Vesico0intestinal fissure. Arch Dis. Child 35:97–102, 1960.
6. Howell C, Caldamone A, Snuder H, et al: Optimal management of cloacal exstrophy. J Pediatr Surg 18:365–369, 1983.
7. Diamond DA, Jeffs RD: Cloacal exstrophy. J Pediatr Surg 18:365–369, 1983.
8. Hurwitz RS, Gian AM, Manzoni AM et al: Cloacal exstrophy: A report of 34 cases. J Urol 138:1060–1064, 1987.
9. Mitchell ME, Brito CG, Rink RC: Cloacal exstrophy reconstruction for urinary continence. J Urol 144–554–558, 1990.
10. Ricketts RR, Woodard JR, Zwiren GT, et al: Modern treatment of cloacal exstrophy. J Pediatr Surg 26:444–450, 1991.
11. Lund DP, Hendren WH: Cloacal exstrophy: A now reconstructible entity. J Pediatr Surg 28:1360, 1993.
12. Hendren WH, Hendren RB: Bladder augmentation: Experience with 129 Children and young adults. J Urol 144:445–453, 1990.
13. Hendren WH: Ileal nipple for continence in cloacal exstropy. J Urol 148:372–379, 1992.
14. Warf BC, Scott RM, Barnes PD et al: Tethered spinal cord in patients with anorectal and urogenital malformations. Pediatr Neurosurg 19:25–30, 1993.
15. Adams MC, Mitchell ME, Rink RC: Gastrocystoplsty: An alternative solution to the problem of urologic reconstruction in the severely compromised patient. J Urol 140:1152–156, 1988.
16. Peters CA, Hendren WH: Splitting the pubis for exposure in difficult reconstructions for incontinence. Urol Clin North Am 17:37–45, 1990.
17. Hendren WH: Management of cloacal malformations. Seminars in Pediatr Surg 6:217–227, 1997.

ABSTRACTS

Abstract No. 1

PHARMACOLOGY, ELECTROPHYSIOLOGY, AND IMMUNOHISTOCHEMISTRY OF THE EXSTROPHIC BLADDER

R. Wammach; M. Fisch, H. Nawrath, E.Weihe, and R. Hohenfellner

University of Mainz School of Medicine, Department of Urology, Pharmacology and Neuroanatomy, Mainz, Germany

Even after multiple reconstructive procedures continence is difficult to achieve in bladder exstrophy patients. Urodynamic investigations have shown that bladder instability and not insufficient bladder outlet resistance may be chiefly responsible for incontinence in these children. It was the aim of this study to investigate whether an inherent pathology of the detrusor muscle of the exstrophy bladder may account for the urodynamic and clinical observations.

Methods: Detrusor muscle specimens were obtained after cystectomy in eight patients with bladder exstrophy. The effects of different drugs were investigated on the spontaneous, pharmacologically and electrically induced contractile activity. Moreover, the patch clamp technique in the whole-cell configuration was used to elucidate the electrophysiologic properties of the exstrophic detrusor muscle cell. Immunohistochemistry was carried out.

Results and Conclusions: All detrusor specimens demonstrated profound spontaneous contractile activity developing a mechanical tension up to 30-fold of the specimens weight. These spontaneous contractions were not influenced by anticholinergic agents. Moreover, a peculiar reaction to acetylcholine was exhibited. The exstrophic bladder was 10 to 100 times more sensitive to acetylcholine than a normal bladder and developed long-lasting, tonic contractions. The quantum of non-adrenergic, non-cholinergic (NANC) neurotransmission was significantly higher in the exstrophy bladder. Consequently, the electrically evoked response could only be minutely influence by oxybutynin. The voltage-clamp experiments showed the exstrophy detrusor muscle cells to have a 10–15 mV higher membrane potential than normal human detrusor cells. A high incidence of spontaneous action potentials was observed. It is concluded that the exstrophic bladder demonstrates unusual physiological and pharmacological properties offering a possible explanation for urodynamic and clinical findings in bladder exstrophy patients.

Abstract No. 2

A MODEL OF ISOLATED BLADDER EXSTROPHY IN THE FETAL LAMB

Leslie D. Tackett, Konstantionos Papadakis, Monique e. DePaepe, Louis J. Wojcik, Francois I. Luks, and Anthony A. Caldamone

Brown University/Hasbro Children's Hospital Departments of Urology and Pediatric Surgery, Providence, Rhode Island

Introduction: The surgical management of bladder exstrophy aimed at normalization of bladder function continues to remain a challenge. Incontinence and bladder dysfunction pose a significant problem for many children with bladder exstrophy. We have created a model of isolated bladder exstrophy in the fetal lamb, without partial bladder wall excision or public bone separation, in an attempt or reproduce the clinical findings (small capacity, noncompliant bladder) seen in exstrophy.

Methods: Fetal lambs at 95–105 days gestation were used. Via an open hysterotomy, the fetal hindlimb was delivered. An infra umbilical ellipse of skin was excised. The bladder was exposed and opened with a Y incision from the dome to the bladder neck. The bladder edges were everted and approximated to the skin edges. The fetal lambs were delivered near term (137 days, term=145 days). At delivery, each exstrophied bladder was closed, a cystometrogram performed, and sections preserved for histologic study. Statistical analysis was performed using the Student's t test.

Results: Twelve (86%) of fourteen lambs were delivered near term with visually successful bladder exstrophy. These twelve bladders were compared with three control nonoperated lamb bladders. Mean bladder weight was 2.56±1.33 g in the exstrophy group vs. 2.82±0.42 g in the controls. Mean fetal weight at delivery and bladder weight to body weight ratio were similar in both groups. Bladder capacity was defined as the maximum bladder volume at a pressure of 10 mmHg. Mean capacity in the exstrophy group was 2.45±1.23 mL and 55.8±6.29 mL in the control bladders ($p<0.0001$). Total compliance was 7.49±4.67 mmHg/mL in the exstrophy group and 0.96±0.58 mmHg/mL in the control group ($p<0.03$). Histologically, the exstrophy bladders demonstrated a thickened submucosa and connective tissue infiltration of the muscle layer relative to the controls.

Conclusions: In this fetal model of isolated bladder exstophy, without pubic diastasis or removal of bladder tissue, the physiologic and histologic changes miror the features of clinical bladder exstrophy. This model should allow the study of the pathogenesis of the small capacity, noncompliant exstrophied bladder, and the development of ante-or postnatal treatment modalities.

Abstract No. 3

INNERVATION PATTERN IN THE EXSTROPY-EPISPADIAS COMPLEX

W.H. Rosch, W.L. Neuhuber*, and K.M. Schrott

Department of Urology, University of Erlangen-Nuernberg, *Institute of Anatomy 1, University of Erlangen-Nuernberg, Germany

While a primary neurogenic dysfunction in exstrophy-epispadias-complex is often postulated, results of urodynamic examination after functional reconstruction in our own Erlangen-technique did not provide any indication for idiopathic dysinnervation. Since there are only few data available on this issue, we decided to apply immuno-cytochemistry for neuronal markers as a first step to elucidate this problem. Transmural biopsies were obtained during reconstructive surgery for exstrophy in 22 cases. Biopsies from 6 patients without bladder malformations who underwent surgery for other reasons were taken as controls. Indirect immunocytochemistry for PGP 9.5, NPY, Substance P, CGRP, VIP and histochemistry for NADPH-d was performed on 14 μm cryostat sections. Slides were evaluated using a semiquantitative approach by three independent investigators. In addition, urodynamic follow-up studies were done 6 months after reconstruction to search for a possible correlation between morphological innervation status and function parameters.

Untreated exstrophic bladders showed an innervation pattern comparable to normal urinary bladder. Some exstrophic bladders after failed reconstruction showed less dense innervation in smooth muscle, but increased innervation in epithelium and lamina propria. Cloacal exstrophies and so-called "transition forms"—exstrophies combined with an anal atresia or a caudal-regression-syndrome—were conspicuous because of extremely dense and irregular innervation.

Morphological data on the innervation pattern concurred well with results of uro-dynamic examination. Patients with "common" exstrophy performed sufficiently well in uro-dynamic follow-up studies indicating successful preservation of nervous structures. Patient after failed reconstruction frequently showed bladder-instability and low-compliance. Patients with cloacal exstrophy and transition forms had severely reduced bladder compliance and small capacity.

These results indicate that "common" exstrophic bladders receive structurally and by inference also functionally intact innervation. Bladders with cloacal exstrophy and "transition forms" display severe irregular hyper innervation which may contribute to the dissatisfying functional outcome of reconstructive surgery.

Abstract No. 4

FEMALE EPISPADIAS REPAIR

P. Dik and T.P.V.M. de Jong

University Hospital for Children and Youth, Utrecht, The Netherlands

Aim: To demonstrate our newly developed technique of female epispadias repair.

Introduction: Female epispadias repair is a challenging operation that should lead to continence and normal cosmetic appearance. No urethra is present in these patients and the insufficient bladder neck is located at the level of the hymenal ring. Correction, therefore, must consist of construction of a neo-urethra and movement of bladder neck and proximal urethra into an intra-abdominal position. We have developed an operation technique that has been very successful in three consecutive cases over the last four years.

Method: The video shows our approach to these patients. The urethral plate is isolated and tubularised. The bladder neck and urethra are liberated from all attachments and pulled p with a modified needle-suspension technique. By doing this, the reconstructed urethra and bladder neck are pulled up as far as necessary to position the neo-meatus at the level of the hymenal ring. After closure of the pelvic floor the clitoris and labia are reconstructed to a normal female appearance.

Results: All patients are able to micturate without residue. Two patients with small bladder capacity needed additional macroplastique injection in the bladder neck to achieve continency, one patient with normal capacity appears to have normal continency in one operation. Cosmetic appearance is completely normal in all three patients.

Conclusion: The demonstrated technique is superior to conventional staged operations for epispadias with standard bladder neck reconstructions.

Abstract No. 5

COMPLETE PENILE DISASSEMBLY TECHNIQUE IN EPISPADIAS REPAIR

Perovic Sava and Djordjevic Miroslav

Department of Urology, University Children's Hospital, Belgrade, Yugoslavia

Aim: Epispadias repair in male remains a great challenge. We report our experience in epispadias repair using Mitchell's technique and its modification.

Materials and Methods: The surgical technique involves complete disassembly of the penis described by Mitchell. The modification is characterized by the preservation of a small tissue "bridge" between the hemiglanses in order to provide better blood supply of the hemiglanses. This procedure does not affect complete dorsal placement of the corporal bodies. During the period from October 1995 to September 1996 the technique was performed on seven patients, aged 2–14 years.

Results: Follow-up was from 2 to 11 months. Dorsal curvature was well corrected in all cases. Penile lengthening was also achieved. There was one meatal stenosis. Cosmetic appearance was good.

Conclusion: Penile complete disassembling presents an improvement in epispadias repair. "Normal" anatomical relations of the penis are achieved by this technique.

Abstract No. 6

RESULTS OF SPHINCTEROPLASTY FOR BLADDER EXSTROPHY

Moneer K. Hanna and Jeffrey A. Stock

New York Hospital-Cornell Medical Center, New York, and Children's Hospital, New Jersey, Newark, New Jersey

Thirty-two children born with classic bladder exstrophy completed staged reconstruction and are now available for evaluation of the urinary tract.

The protocol for functional reconstruction was: bladder closure soon after birth or as early as possible upon referral. Bilateral iliac osteotomy was performed in 14 children and pubic osteotomy in 2. At one year of age the epispadias was repaired in boys Bladder neck reconstruction and ureteral reimplantation were performed after 5 years of age when the bladder capacity under anesthesia was 90–100ml. The results are as follows:

Primary repair	Augmentation	Result	Secondary repair	Result
Young-Dees (15)	4	8 Dry & Voiding		
		7 Wet	Redo Young-Dees (1)	Wet
			Gracilis Ms. (1)	Dry
A.U.S. (4)	2	1 Erosion	Continent Div. (6)	Dry
		3 Late Failure	Continent Div. (2)	Dry
Dry			Lost to F.U. (1)	
Gracilus Ms. (2)	2	2 Dry		
The Cinch (11)	4	5 Dry & Voiding		
		4 Dry & C.I.C.		
		2 Voiding but damp	Collagen Injection	Improved

2/32 developed bilateral hydronephrosis when they became continent of urine. Both are now on C.I.C. with considerable improvement. In this series the early successes with artificial urinary sphincters proved to be long term failures.

It is concluded that the modified cinch procedure may be the ideal procedure for sphincteroplasty. The circular orientation of the muscle flap provides an efficient, leak proof sphincter mechanism.

The gracilus muscle provides a well vascularized graft to support the fibrotic and often thin walled bladder neck. Furthermore it proved to effectively provide dryness and is an excellent salvage procedure for these difficult cases.

Abstract No. 7

THE "PELVIC BUCKET", AN ADJUNCT TO THE CLOSURE OF BLADDER EXSTROPHY IN THE NEONATE

Mark F. Bellinger, M.D., Morey S. More land, M.D., and Francis X. Schnook, M.D.

The Departments of Urology and Orthopedics, The Children's Hospital of Pittsburgh, The University of Pittsburgh, Pittsburgh, Pennsylvania

Postoperative management of the neonate after closure of bladder exstrophy has as significant impact on the success or failure of primary closure. In particular, successful management of the bony pelvis may diminish the degree of postoperative diastasis of the pubic symphysis and the potential for prolapse or recurrent exstrophy of the bladder.

Many techniques have been described for management of the infant pelvis after primary closure, whether osteotomy is performed or not. These include absence of fixation or positioning, a "mermaid" wrap of the legs with the hips and knees flexed at 90 degrees and lower legs internally rotated, lower extremity traction, and external fixation of the pelvis. Each technique has its advantages and disadvantages, and some (traction) require considerable lengths of post-operative hospitalization (up to three weeks). We have evolved a plan for management in most cases which starts with approximately 10 days of lower extremity traction followed by management in an individually-fitted "pelvic bucket" in which the infant is discharged to home, combined with a loose wrap of the lower legs to minimize motion and extraction of the public symphysis.

The pelvic bucket was used in 4 patients with classical exstrophy and one with cloacal exstrophy. Four of the five were initially treated in modified Bryant's traction for 8 to 14 days before being placed in the bucket. Two patients were removed from the bucket at 8 and 9 days because of skin irritation, in one of the patients because of significant weight gain. Length of hospitalization after primary closure ranged from 8 to 25 days with a mean of 13.6 days. After primary closure, one patient suffered modest bladder prolapse, one female is continent and three are stable pending continence procedures. We believe that use of the pelvic bucket is an adjunct to management of the exstrophic bladder in the neonate which can diminish the length of hospitalization after primary closure.

Abstract No. 8

THE MODERN APPROACH TO EXSTROPHY CLOSURE

John W. Duckett, Jr. MD, Thomas Lanchoney, and Douglas A. Canning, MD

Children's Hospital of Philadelphia, Philadelphia Pennsylvania

A video will be presented demonstrating the newborn closure combining bladder neck reconfiguration and epispadias repair with symphyseal approximation without osteotomies. This technique can be performed in the first 48 hours of life. It combines aspects of the Kelly, Chisholm, and Mitchell repair. The initial exstrophy dissection is carried down through the bladder neck and the distal urethra is mobilized off of the corporal bodies bilaterally. The corporal tissue is then split and the urethral plate is taken on its prostatic pedicle posteriorly to release the bladder neck and prostatic urethra.

We believe that this approach facilitates the placement of the bladder neck and prostatic urethra deep within the levator complex without the aid of paraexstrophy flaps. After this dissection, the bladder is closed in the conventional way but the closure is carried out beyond the bladder neck out to the tip of the urethra. The urethra is placed ventral to the corporal bodies which are rotated to correct the dorsal chordee. Following the complete anatomic closure the pubic bones are brought together on the midline with single #1 nylon suture. The urethral is brought to the tip of the reconstructed penis. The fascial tissue of the abdominal wall and the perineum are then reconstructed in the conventional manner and the ventral foreskin is used to cover the dorsum of the penis as has been previously described.

In this child, a bilateral reimplantation was performed six months later with very good results. The child is now voiding intermittently with a small residual urine.

We believe that this anatomic closure provides the opportunity for the bladder to do volume work against a physiologic resistance through the reconstructed bladder neck. Intraoperative cystoscopic views will be shown in the video that demonstrate what we believe is external sphincter action following the anatomic closure. The effect of the reconstructed bladder neck may be to provide a more physiologic resistance which may stimulate bladder growth and ultimately lead to a higher percentage of patients who are able to void with continence.

Abstract No. 9

BLADDER EXSTROPHY: PRIMARY RECONSTRUCTION AND AUGMENTATION WITH DEMUSCOSALIZD COLON

Salvador, V.C. Lima, Luiz Alberto P. Araujo, Marisa Montoro and Fabio O.

Vilar. Hospital das Clinicas, Federal University of Pernambuco and Hospital Infantil Manoel Almeida. Reci-fe-Brazil

Objective - To offer a new way of management of this complex anomaly.

Patients and Methodes - Eighteen patients, 13 males and 5 females with ages ranging from 3 months to 32 years had their bladders augmented or reconstructed during the last 33 months. Fourteen patients had the bladder augmented and 4 used the same procedure in primary reconstruction. Five patients had the deepithelialized sigmoid colon applied over the bladder with the urine in contact with the raw surface. The remaining had a silicone bubble filled with saline inside the reconstructed bladder for 2 weeks in order to keep the patch under some distension.

Results - Follow-up ranged from 1 to 33 months (mean 15.6 median 17). Patients where the silicone model was not used the bladder did not achieve improvement in capacity. In the second group there was significant improvement. Four patients are continent, 5 patients cannot be evaluated and 4 are incontinent.

Conclusion: The use of demucosalized sigmoid colon over a silicone model may represent a feasible method in dealing with this complex anomaly.

Abstract No. 10

SEMEN ANALYSIS AND TRANSRECTAL ULTRASOUND FINDINGS IN MEN WITH BLADDER EXSTROPHY

Terry W. Hensle MD, Erik T. Goluboff MD and Haryy Fisch MD

Dept. Of Urology, Babies and Children's Hospital, Columbia-Presbyterian Medical Center, New York, New York

Althought it has been reported that women with bladder exstrophy have relatively normal potential for fertiilty, little has been reported about such potential in men with this disorder. In 1991, Hamper et al evaluated 7 men with bladder exstrophy with transrectal ultrasound (TRUS) and found that, although 6 had prostates and seminal vesicles, only 2 showed a normal appearance of these structures. In 1979, Lattimer et al reported on semen analysis from 9 men with bladder exstrophy and found 7 were oligo- or azoospermic. Since no study to date has simultaneously examined semen analyses, TRUS findings and hormonal parameters in men with bladder exstrophy, we gathered this data from 4 men with this disorder. The results are as follows:

Pt#	Age	Fert	Atrophy	FSH	Test	Vol	pH	Dens	Motil	Prostate	SVs
1	34	y	Left	5	523	1.5	8.5	203	40%	13cc	low
2	32	N	right	12	535	0.5	8.5	1.5	20%	15cc	low, dilated
3	26	N	N	16	314	0.3	8.5	1.5	30%	36cc	normal
4	24	N	Right	30	335	0.3	8.5	0	0	cystic	low, dilated

(Age in years, Fert=known fertility, Atrophy=testicular atrophy, FSH=Follicle stimulating hormone concentration in mIU/ml. Test=testosterone level in ng/dl, vol=semen volume in cc, dens=sperm density in 10^6ml., pH=semen pH, Motil=sperm motility in %, prostate, Svs=TRUS appearance of prostate and seminal vesicles. Low Svs=abnormal distal insertion of seminal vesicle/ejaculatory ducts into the prostate)

In summary, FSH concentrations were normal in 2 of 4 patients while testosterone levels were all normal. Semen analyses were significant for a pH greater than 8.5 and a low semen volume in all patients. Sperm density and motility ranged from aspermic to normal. TRUS findings were significant for prostates that ranged from cystic to small to normal. Most patients had abnormal distal insertion of the seminal vesicles/ejaculatory ducts into the prostate. In conclusion, fertility potential in male exstrophy patients appears variable. Three possible explanations for these findings are: 1) a developmental defect in this disorder may involve hindered growth of prostatic tissue acounting for the elevated seminal pH, low semen volume and abnormal seminal vesicle-prostatic insertion that appears to be present in these patients. 2) alterations in anatomy could be secondary to surgical procedures undergone by these patients and 3) functional changes in the prostate could result from multiple episodes of epididymitis.

Abstract No. 11

DUPLICATE BLADDER EXSTROPHY: A UNIQUE APPROACH TO INITIAL REPAIR

Stephen Strup, Michael Fabrizio, R. Bruce Filmer, John Noseworthy, and John P. Gearhart**

Philadelphia, Pennsylvania and **Baltimore, Maryland (Presented by Dr. Fabrizio)

Introduction: Bladder exstrophy has been successfully managed by staged surgical repair with early bladder closure, subsequent epispadias repair and finally, bladder neck reconstruction. Duplicate bladder exstrophy is a rare variant of the exstrophy complex with fewer than twenty cases reported. We describe a unique approach to the management of duplicate bladder exstrophy combining initial bladder closure and epispadias repair.

Materials and Methods: A male newborn presented with the appearance of classic bladder exstrophy and a ruptured omphalocele. Repair of the omphalocele and bilateral orchidopexy was performed shortly after birth but bladder closure was delayed due to his very small bladder. At age 8 months, bladder closure was performed with intraoperative finding of a duplicate bladder lying posterior to the exstrophied bladder. The left ureter drained into the superior bladder and the right ureter drained into the posterior bladder. After anterior and posterior osteotomies, the left ureter was reimplanted in to the posterior bladder. A portion of the exstrophied bladder was then used to construct a neourethra which was used for the concurrent epispadias repiar.

Results: The child has done well with an excellent cosmetic result. The bladder subsequently required augmentation which was done at the time of his Youg-Dees-Leadbetter bladder neck reconstruction (age 3 1/2). The child is currently continent but does require intermittent catheterization through an appendicovesicostomy.

Conclusions: Our unique surgical approach is an option for successful management of this rare variant.

Abstract No. 12

THE EXSTROPHY AND EPISPADIAS COMPLEX IN THE DUPLICATED LOWER URINARY TRACT

Fabienne Perren and Peter Frey

Dept. of Pediatric Surgery, Centre Hospitalier Univesitaire Vaudois, Lausanne, Switzerland

Purpose: The exstrophy-epispadias complex in combination with a duplicated lower urinary tract is an extremely rare congenital malformation. It can, in addition, involve complex malformations of the urinary, genital, digestive and musculoskeletal systems. The aim of our paper is to describe two such cases of bladder exstrophy in association with duplicated lower urinary tract.

Material and Methods: The first case was and 8 months-old baby presenting an anterio-posterior duplication of the bladder with a communicating isthmus and a duplicated urethra. The anterior system was exsrophic and epispadiac.

The second case was a female newborn with duplicated bladder without epispadias. There was communication betwen the anterior exstrophic bladder and the posterior normal bladder. A careful examination revealed no other malfomations.

Results: In the first case the anterior exstrophic bladder was resected just anterior to the isthmus and the posterior bladder was then closed. The epispadiac urethral plate was then removed, the glans reconstructed and the penile skin closed dorsally. Tension-free closure of the abdominal wall was achieved. The post-operative recovery was uneventful and full urinary continence was present. The diagnosis was confirmed histologically.

In the second case, the anterior part of the bladder was resected above the isthmus. Bladder mucosa and detrusor muscle were closed in layers. The abdominal wall was reconstructed. Recovery was uncomplicated, the child showed normal voiding. Histological examination confirmed the diagnosed bladder duplication.

Conclusion: Although duplication of the lower urinary tract in combination with exstrophy and epispadias complex is extremely rare, all children presenting with an exstrophic bladder and a normal urethra alone or in association with an epispadias anterior urethra should be investigated as regards to bladder duplication with possible isthmic communication.

Abstract No. 13

FOLLOW UP OF FIVE YEARS AND MORE AFTER ANTERIOR PELVIC OSTEOTOMY IN BLADDER EXSTROPHY REPAIR

Perovic Sava and Djordjevic Miroslav

Department of Urology, University Children's Hospital, Belgrade, Yugoslavia

AIM: We report postoperative results 5 years or more after anterior pelvic osteotomy in patients with bladder exstrophy.

MATERIALS AND METHOD: During the period from August 1988 to December 1991, 36 patients aged from 3 months to 12 years underwent surgery. The operative technique involved bilateral osteotomy of the superior ramus of the pubic bone described by Frey. In small children ischiopubic junction allowed symphyseal approximation, while in older children this was achieved by fracture of interior ramus of the pubic bone. A combination of wires, bolts and plates is used for osteosynthesis. The bladder was either cloned or, in most cases, the exstrophic bladder plate was inserted deep into the pelvis, allowing subsequent epithelialization of the bladder and further formation and growth.

RESULTS: Follow-up was from 5 to 8.5 years (mean 6.5 years). A stable pelvic ring without compromised gait was achieved in all patients. Symphyseal rediastasis occurred in 8 patients; complete in one and moderate in 7 patients, but without affecting the final result of reconstruction of anterior abdominal wall and bladder. Four children developed a hernia. Primary bladder reconstruction was successful and bladder capacity was from 50 to 200 ml.

CONCLUSION: Our postoperative results after 5 years and more follow-up prove that anterior pelvic osteotomy could be a good alternative to other pelvic osteotomies.

Abstract No. 14

BILATERAL POSTERIOR PELVIC RESECTION OSTEOTOMIES FOR EXSTROPHY OF THE BLADDER

J. J. Gugenheim, M.D. (Texas Orthopedic Hospital, Houston, Texas 77030), E. T. Gonzales, Jr., M.D., D. R. Roth, M.D., and B. A. Montagnino, R.N., M.S.

Texas Children's Hospital, Houston, Texas

We describe a modification of bilateral posterior iliac osteotomies for bladder exstrophy, in which a strip of ilium is resected subperiosteally lateral to the sacroiliac joints, allowing easier anterior closure with less breakdown compared to traditional osteotomies. Thirty-one children underwent repair of bladder exstrophy between 1974 and 1994. Bladder procedures included: closed reduction and cast application in the newborn period (4), classic bilateral posterior iliac osteotomies (12), and bilateral posterior resection osteotomies (15). Dehiscence occurred following one closed reduction, five classic osteotomies, and one resection osteotomy. Urinary continence was obtained in four cases of closed reduction, nine cases of classic posterior osteotomies, and nine cases of posterior resection osteotomies.

This procedure offers several advantages: The transverse incisions are in Langer's lines and heal with excellent cosmetic results. It can be done at any age, whereas it is difficult to use an external fixator for a very small child or obtain adequate mobility with traditional osteotomies on an older child. The incisions are not near ostomy sites for a patient with cloacal exstrophy.

Abstract No. 15

BIOMECHANICAL ANALYSIS OF HIPS IN ADULTS WITH BLADDER EXSTROPHY

Mihir M. Jani, Paul D. Sponseller, M.D., John P. Gearhart, M.D., and Robert D. Jeffs, M.D.

Baltimore, Maryland

Introduction: Children and adults with bladder exstrophy have an open pelvic ring. Although the symphyseal diastasis is reduced with osteotomy, it increases with age as a part of normal growth. The natural history of the hip joint with an open pelvic ring has not been studied. Understanding the biomechanics of the hip in bladder exstrophy may indicate whether exstrophy patients are at greater risk for degenerative joint disease (DJD).

Materials and Methods: We studied 14 patients 16 to 52 years of age born with classic bladder exstrophy. Twelve patients did not have pelvic osteotomies at exstrophy closure whereas 2 patients had osteotomies with essentially no correction. Patients were matched for age and gender with 12 controls who had normal AP pelvis radiographs. AP pelvis radiographs were used for biomechanical analysis which was based on the single-stance phase of gait at which the maximum load and stress is exerted on the hip joint. Joint force and joint stress (force/area) were calculated relative to partial body weight by the method of Legal (1987), which allowed comparison between patients and with the controls. Statistical analysis was done using several t tests and the level of significance was $p<0.05$.

Results: The mean relative joint force and stress for controls was 3.04±0.26 and 0.177±0.035 cm^{-2} respectively. For the exstrophy patients, the mean relative joint force (4.24±0.91) was significantly higher ($p<0.01$). The mean relative joint stress (0.298±0.18 cm^{-2}) for exstrophy patients was significantly higher as well ($p<0.05$). In addition, the mean distance from the body center to the center of the femoral head (*d6)* was significantly increased ($p<0.001$) in bladder exstrophy patients (12.35 ±1.05 cm) versus controls (10.31±0.70 cm). An increase in diastasis correlated with an approximate 30% increase in the distance *d6.* The mean distance from the greater trochanter to the femoral head center (*t')* was significantly less ($p<0.02$) for exstrophy patients (5.21±1.24cm) compared to controls (6.52±0.80 cm). The center-edge angle (CE angle) was significantly decreased in exstrophy patients (25.8±9.9 deg.) Versus controls (33.3 ±7.73 deg.)($p<0.05$).

Conclusions: The force and stress on the hip joint are increased in adult bladder exstrophy patients. This is due to a significant increase in symphyseal diastasis and thus an increase in the distance *d6* in exstrophy patients which correlates with an almost linear increase in joint load and stress (Legal 1987). The significant decreases in *t'* and the CE angle contribute to increased load and stress as well. Increased joint load and stress can lead to an earlier onset of DJD. Attempts at closing the pelvic ring at exstrophy closure may help reduce the potential maximum diastasis in life and thus reduce the distance from the body midline to the hip joint. This would lead to a reduction in potential joint load and stress and reduce the risk of DJD.

Abstract No. 16

TOMODENSITOMETRY AND OSTEOTOMY OF THE PELVIS IN THE EXSTROPHY COMPLEX

Alain Wakim*, MD, Lotfi Miladi*, MD, Jean-Patrick Barbet*, MD,
Pierre-George Helardot*, MD, and Jean Dubousset*, MD

*Department of Pediatric Surgery, **Department of Pathology, Hopital Saint Vincent de Paul, 82 avenue Denfert-Rochereau, 75674 Paris, France

By using computed tomography scanner (CT Scan) of the pelvis of fetuses with bladder exstrophy, we have elaborated a new method of investigation of the anomaly. The tomography slices were selected at the same levels for each case. Three angles expressing external opening and widening of the anterior pelvis were defined.

Comparing normal and abnormal pelvises allow to define three criterias for the correction of the malformation: 1/ the sum of the differential angles gives the amplitude of the correction needed; 2/ a supra-acetabular osteotomy appears to allow best closure of the pelvic ring; 3/ only 3 slices of CT Scan are needed to define these criterias, allowing its use in neonates or infants.

We received 16 children (14 with bladder exstrophy and 2 with epispadias) operated between June 1992 and January 1997. For 12 of them, a CT Scan was performed 1 to 4 years after osteotomy of the pelvis. For the remaining 4 patients (2 with bladder exstrophy and 2 with epispadias) a CT Scan was performed before and, in one case, one year after the reconstructive surgery. Results showed antero-posterior diameter of the pelvis around 30% shorter than in normal pelvis, and increasing shortening of nearly 30% with closure of the pubic symphysis by osteotomy. But this diameter was nearly normal and associated with very good continence in 5 out of 8 patients in whom osteotomy had been performed in the first days of life or in infancy.

In conclusion:

- a CT Scan of the pelvis should be performed every time an osteotomy of the pelvis is planned.
- the antero-posterior diameter of the pelvis seems to be a predictive criteria to future continence.
- in failed exstrophy patients, results of the reconstructive surgery with osteotomy could be eventually improved by distraction of t he anterior segment of the pelvic ring.

Abstract No. 17

GONADAL MORPHOLOGY IN CLOACAL EXSTROPHY: IMPLICATIONS IN GENDER ASSIGNMENT

David W. Marsh, Elizabeth J. Perlman, Ranjiv Mathews, Robert Jeffs, and John P. Gearhart

Baltimore, Maryland (Presentation by Ranjiv Mathews)

INTRODUCTION: Cloacal exstrophy is the most severe variant of the exstrophy-epispadias complex. Associated defects of the neural axis and digestive tract are common. The extensive urological deformity makes management difficult. In genotypic males, early gonadectomy and gender reassignment has been performed because of difficulty in surgical reconstruction of a "functional" male. The extensive nature of the morphogenic defect could involve gonadal histology.

METHODS: Gonadal histology of fifteen genotypic males who had gonadectomy and gender reassignment, were evaluated for the presence of histologic abnormalities. Age at gonadectomy ranged from 0–20 months (mean—6 months). Age matched normal autopsy controls were used for comparison.

RESULTS: Architecture of the rete testes, epididymis and vasa of all specimens were consistent with normal infantile tissue. Some of the study specimens demonstrated thinning of the tunica albuginea with a persistent cellular layer. This feature has also been noted in normal prepubertal testes. Seminiferous tubular histology was normal except for a qualitative increase in the proportion of germ cells. The increase in the proportion of germ cells was independent of the interval to gonadectomy.

CONCLUSION: Despite the severity of the defect in cloacal exstrophy, gonadal histology is preserved. In those patients raised as males, gonadal preservation with early orchidopexy should be performed. The significance of the increase in germ cell numbers is unknown and deserves further evaluation.

Abstract No. 18

CLOACAL EXSTROPHY: EVALUATION OF URINARY CONTINENCE FOLLOWING STAGED BLADDER RECONSTRUCTION

Douglas A. Husmann,[1] David R. Vandersteen,[2] Gordon A. McLorie,[3] and Bernard M. Churchill[3]

[1]Rochester, Minnesota and Dallas, Texas, [2]Rochester, Mimmesota, [3]Toronto, Ontario, Canada

PURPOSE: Reconstruction of the bladder in cloacal exstrophy is a difficult technical challenge. We reviewed our patient populations with cloacal exstrophy that were repaired with a staged bladder reconstruction. Specifically our goal was to observe the incidence of urinary continence following Young Dees Leadbetter reconstruction of the bladder neck in cloacal exstrophy.

MATERIALS AND METHODS: We received all patients with cloacal exstrophy whose bladders were managed by staged reconstruction from 1971–1976. All of the patients within this study had a Young Dees Leadbetter bladder neck reconstruction as their second stage procedure. A control population of patients with classical bladder exstrophy managed in a similar fashion were used as controls.

RESULTS: Twenty-three patients with cloacal exstrophy completed their two stage repair. Five (22%) are continent, 3 of whom can not void and use intermittent catheterization to empty. Eighteen patients (78%) remained incontinent following their staged reconstruction. Six (26%) eventually became continent by surgical revision using a Mitrofanoff principle. Five (22%) chose to have a urinary conduit performed rather than proceed with attempted continent urinary reconstructions. Seven (30%) have persistent incontinence per urethra, two of whom failed attempted AUS placement (both eroded). In contrast to the cloacal exstrophy patient, staged reconstruction in classical bladder exstrophy resulted - 67/82 (82%) of our patients becoming continent. Twenty-three required intermittent catheterization to empty.

CONCLUSION: The ability of the Young Dees Leadbetter bladder neck reconstruction to result in urinary continence is significantly different between the cloacal and classical bladder exstrophy patient populations, 22% vs. 82% respectively, $p<0.001$. This finding results in our suggestion that alternative surgical methods other than a Young Dees Leadbetter bladder neck procedure be considered when planning for a urinary continent procedure in children with cloacal bladder exstrophy.

INDEX